The 5 Laws of Applied Intelligence

Why Your Humanity Is Your Superpower and How to Use It to Win in a World of Machines

Angela Radcliffe

❀ Formatted with Vellum

For my children, and yours. May we leave them a world where technology amplifies humanity rather than replacing it, where agency is accessible to everyone, and where the wisdom we pass on empowers them to build things that matter.

Foreword: Inheritance

Both Inheritances

My father once spent twenty minutes arguing with me about which AMC theater was next to the Burlington Coat Factory in our local mall.

I was fifteen. I was absolutely certain I was right because I'd spent countless hours at Beltway Plaza Mall. It was my teenage refuge, the place where I could window-shop dreams I couldn't afford. I knew every store, every food court vendor, every shortcut through the parking lot. And on this particular occasion, I cared very much about which theater was where, because I was meeting a boy my father absolutely did not know about. If he picked me up at the wrong anchor store, my cover was blown.

But my father? He'd maybe visited twice in his life. Yet there he stood in our living room, carpenter's pencil in hand, sketching his version of the mall layout directly onto a piece of drywall he'd been meaning to hang for weeks. Because of course we had drywall leaning against the living room wall. Duct tape, chicken wire, plywood, and the occasional power tool were in common supply in our falling-down house.

"It's right here," he said, tapping the diagram. "AMC 8, next to Burlington."

"Dad, I *go* there. Like, all the time. That's not where it is."

"The facts are the facts," he said, and drew another line.

I was the oldest of six kids, which meant every disagreement became a spectator sport. Within minutes, we had an audience. Voices were raised. The drywall stayed against the wall for weeks after, permanently marked with our competing versions of reality. A monument to our mutual certainty.

What neither of us knew was that there were two AMC theaters in that mall. I was talking about one. He was talking about the other. We were both right. We were both wrong. We were both so convinced of our own perspective that we couldn't see the other person was operating from an entirely different, and equally valid, piece of truth.

I think about that argument constantly, thirty-six years later. Because it's exactly what's happening in every conversation about artificial intelligence.

My father kept a copy of the Federalist Papers in his truck cab. Not exactly beach reading between deliveries. He came from a long line of writers and editors. He had a nearly photographic memory and never let us use incorrect grammar. He understood something fundamental: we may have been poor, but we had the wealth of intelligence.

"Intelligence is the only inheritance that compounds over time," he told me more than once. "Everything else can be taken from you. Your house, your job, your health, your savings. But the ability to think critically, to make your case, to parse complexity into clarity? That's yours forever."

That inheritance changed my trajectory. The ability to make an argument, to claim intellectual ground even when systems said I didn't belong there, to trust that my intelligence mattered regardless of credentials. A girl with no college degree, no connections, no safety net. Just a quick wit, a loud mouth, and a father who taught me that you don't need a seat at the table if you know how to build your own.

That's the first inheritance this book offers you: *permission to be in the conversation.*

It took me far too long to learn that vision without execution is just daydreaming.

My mother never read the Federalist Papers. She was too busy keeping six kids alive on peanut butter sandwiches and Imperial margarine, serving meals on unbreakable Corelle plates with mustard-colored flowers around the rim. She doesn't theorize about gardening. She just gardened.

When my brother Levi was a little boy and decided he wanted to grow a grapefruit tree from a seed, my mother handed him an old chipped coffee mug and a spoon.

"Go dig in the dirt," she said.

She didn't expect it to work. But she understood something my father's intellectual inheritance couldn't teach me: *doing beats theorizing every single time.*

That grapefruit tree grew. And grew. And grew until it was large enough that it had to live inside during Maryland winters, moving from house to house with us, a silent witness to our chaotic family. It was there in the corner the morning the soldiers came to tell my parents that Levi had died. Twenty-one years old. An undiagnosed heart condition. The tree outlived the boy who planted it.

My mother always said she'd write a story about it someday. From the tree's perspective, watching us all grow up in our loud, laughing, yelling way. She never did write that story. She was too busy actually living it.

If my father taught me that intelligence is the only inheritance that compounds over time, my mother taught me the corollary: *intelligence that doesn't do anything is just performance.*

My father gave me permission to claim intellectual ground.

My mother gave me the spine to hold it.

What I've learned after twenty-five years in healthcare innovation, data governance, clinical research, and AI literacy education is that you need both. Especially right now.

This is what I mean by Applied Intelligence.

A Note on What's at Stake

I need to be honest with you before we go any further.

AI is not coming. AI is here. It is reshaping industries faster than most people can comprehend. In the pharmaceutical world where I spent much of my career, job tenures that were fifteen to twenty years just a few years ago have shrunk to three to five. Now I'm seeing twelve to eighteen months. That's not a prediction. That's what's already happening.

And it's not just jobs. The systems being built right now are trained on the labor of workers in the Global South who are paid pennies to label data so the models can learn. The energy required to run these systems is staggering. And yes, your hour of streaming Netflix or your long hot shower is part of that same climate equation, so let's be honest about all of it. The biases baked into AI reflect every inequality we've ever encoded into data, racial, economic, gendered, and those biases don't just persist. They scale.

I'm not telling you this to scare you. I think you're already scared. Or overwhelmed. Or exhausted by people telling you to be scared.

I'm telling you this because the biggest danger isn't AI itself. The biggest danger is smart, capable people, people like you, sitting on the sidelines while a handful of companies and technologists make decisions that will shape your life, your children's education, your healthcare, your career, and your community. They are not thinking about you when they make those decisions. I promise you that.

This book is not Pollyanna. Applied Intelligence is not optimism. It's pragmatism with a backbone.

Who This Book Is For

This book is for people who have deep expertise in their domains, in their work, their communities, their families, but have been told they don't belong in conversations about AI because they're "not technical enough."

It's for the nurse with twenty years of bedside experience who's been told she doesn't understand AI. It's for the teacher who knows her students better than any algorithm but feels unqualified to question edtech tools. It's for the parent who can't code but can absolutely tell when something isn't serving their child's development.

But it's also for the people who don't fall neatly into the camps of fear or hype. The ones who can see the potential. Who aren't rejecting AI and aren't blindly adopting it, but who still don't quite grasp how fast the ground is shifting under their feet. You're not panicking. But you're not moving, either. And right now, standing still is its own kind of risk.

If you have expertise in anything, if you have lived experience, contextual knowledge, human judgment, you belong in this conversation. Not someday. Now.

I'm not teaching you everything you need to know. I'm giving you the confidence that you can figure it out, and a framework for beginning. With your eyes wide open.

The door isn't as locked as they want you to believe. And you already have everything you need to open it.

Welcome to Applied Intelligence.

The Two Theaters

On Perspective, Expertise, and the Trap That Keeps Smart People Silent

What that ridiculous teenage argument taught me is this: **you don't need to see the whole mall to know your section matters.**

My father had barely visited Beltway Plaza, but his diagram of his section was accurate. I practically lived there, but I'd never been to the section he was describing. We both had legitimate knowledge. We both had real expertise. We were both qualified to speak about what we knew.

The problem wasn't that one of us lacked credibility. The problem was that we each assumed our limited perspective represented the complete picture.

This is happening right now in every conversation about AI. The technologists see one theater. The ethicists see another. The executives see a third. The workers whose jobs are changing see a fourth. Everyone is certain. Almost no one is asking what the person in the next section of the mall can see that they can't.

And the people with the most at stake, the ones living closest to the actual problems, are often the last to be asked.

Proximity Creates Expertise

Whether it's being said out loud or not by those with credentials and those considered experts, the most valuable form of expertise in any system is **contextual knowledge.**

Contextual knowledge is knowing that the third printer on the second floor always jams, so you send important documents to the first-floor printer even though it's farther away.

Contextual knowledge is understanding that Dr. Peterson is always running thirty minutes behind on Thursday afternoons, so you don't schedule difficult patient conversations then.

Contextual knowledge is recognizing that your daughter focuses better on homework after she's had twenty minutes of physical activity.

This is the knowledge that comes from proximity. From actually living in the system. From being there, day after day, watching what works and what doesn't.

Let me give you an example from my own work.

When I was leading data projects at a global biopharmaceutical company, we were implementing a new data governance system. The consultant team we'd hired had impressive credentials. Multiple PhDs. Years of experience. Fancy frameworks with names that sounded important in slide decks. They were building something elegant. The problem was, nobody was going to use it.

They kept running into the same wall. The early discovery scientists didn't trust the late-stage clinical development scientists with their data, and vice versa. Both sides had legitimate concerns. Both sides were also dug in. The consultants tried to solve it with architecture and governance models. It wasn't working.

The breakthrough came when I stopped trying to get the consultants to figure it out and started sitting between the two groups myself. I didn't have their technical depth. I couldn't have designed the database architecture. But I could listen to what each side was actually worried about, underneath the jargon and the posturing, and broker an agreement by addressing their practical concerns about what it would mean if data was shared across stages.

I could tell you precisely what needed to happen for the system to actually serve the people using it, because I was close enough to the problem. Closer than anyone with a PhD in the room.

That's Applied Intelligence. Not replacing technical expertise, but recognizing that technical expertise alone is insufficient. The person closest to the problem often holds the piece that makes the solution actually work.

The False Dichotomy

I've spent a significant part of my career advocating for equitable representation of patients in clinical research and for access to healthcare. One of the things that work taught me is that the people who navigate complex systems most effectively are rarely the ones with the most technical knowledge. They're the ones with the most proximity to the problem.

Think about what happens when someone you love gets seriously ill. The healthcare system is staggeringly complex. Insurance networks, specialist referrals, treatment protocols, prior authorizations, clinical trial eligibility, formulary restrictions. You don't need a medical degree to know that your father's oncologist isn't communicating with his cardiologist. You don't need to understand hospital billing software to know that the bill you just received doesn't match what you were told. You don't need a background in clinical research to know that your child's symptoms aren't being taken seriously.

I've watched patient advocates navigate these systems with extraordinary effectiveness. Not because they understand the technical infrastructure of healthcare IT. Because they understand where the friction is. Where the data gets stuck. Where the processes break down. Where a human being needs to step in because the system isn't designed to handle their specific situation.

That contextual knowledge, built through proximity and lived experience, is the same kind of expertise that matters in conversations about AI.

And yet. When the conversation turns to AI, the same people who can navigate a byzantine healthcare system on behalf of someone they love suddenly go quiet. "I'm not technical," they say. "I don't understand how it works." As if understanding the algorithm is a prerequisite for having an opinion about what it should do.

This is the false dichotomy that's paralyzing an entire generation of smart people: either you're a technical expert who understands AI at the algorithmic level, or you're obsolete.

It's a lie. And it's a lie that benefits the people selling you the tools.

You don't need to understand how the algorithm works to know when something isn't serving the people it's supposed to serve. You've been doing that your whole life. In doctor's offices, in school board meetings, in workplaces where somebody made a decision that looked right on a spreadsheet but wrong in practice.

Your perspective isn't just valid. It's the piece that's missing from almost every AI implementation I've ever seen fail.

The Intelligence Trap

So if your perspective matters this much, why are so many smart people surrendering it?

I used to work for a programmatic media company. If you're not sure what that means, think of it this way: it's the technology behind that moment when you mention binoculars for birdwatching over dinner and then an ad for binoculars shows up on your phone twenty minutes later. Programmatic media places highly tailored ads in front of you based on the context it can see. It was the beginning of personalized advertising, and at the time, it felt like magic.

I sat in a lot of rooms where it felt like magic to the clients, too. They would watch the demos, hear the terminology, see the dashboards, and their eyes would glaze over with a very specific kind of awe. Knowledge graphs. Real-time bidding. Contextual targeting. The jargon piled up and the clients nodded along.

These were smart people. Marketing directors, brand leaders, people who had spent years, sometimes decades, understanding their audiences. They knew who their customers were, what motivated them, what language landed and what fell flat. They held deep, earned expertise about the humans on the other end of those campaigns.

But the second the conversation turned technical, they went quiet. They stopped asking whether programmatic was the *right* strategy for their audience and started assuming it was the *best* strategy, because the people presenting it seemed so sure. Even when a higher-touch, more human approach to promotion would have made more sense for their specific situation, they deferred. Not because they lacked judgment. Because the room made them feel like their judgment didn't count.

I watched this happen over and over. And I see the exact same thing happening now with AI, at a much larger scale and with much higher stakes.

This is the intelligence trap.

The Two Camps

In every conversation I have about AI, I see people splitting into two camps. Both camps are responding to the same anxiety. Both share the same fundamental assumption. And both are wrong.

The Surrenderers believe big tech has already won. The systems are too complex. The companies too powerful. The trajectory too inevitable. So they yield. They adopt every new tool without question. They defer to the algorithm.

You've seen this. The executive who lets IT choose an enterprise software platform without asking what the downstream implications will be for the people who actually have to use it every day. The hiring manager who trusts the AI screening tool to eliminate candidates without reviewing who got filtered out. The parent who hands their kid a tablet loaded with AI-powered educational apps and assumes the technology knows better than they do.

Surrenderers aren't lazy. They're overwhelmed. They've been told, in boardrooms and on stages and in breathless LinkedIn posts, that AI is smarter, faster, and more objective than they could ever be.

The Resisters believe the only moral choice is rejection. They see surveillance. Manipulation. Job loss. Dehumanization. So they fight. They ban AI in their classrooms. They refuse it in their businesses. They avoid the tools entirely.

I recently worked with a videographer. Talented guy. He knew how to frame a shot, had impressive studio space, and could direct me in front of a camera better than anyone I've worked with. We were filming content about AI, which made it all the more striking when he told me, point blank, that he was anti-AI and believed it was going to destroy all creative work.

At the same time, he mentioned that he was trying to buy out a business partner, struggling with scheduling, and looking for ways to scale his studio. All problems where AI tools could make a real difference. But he'd built such a hard wall against it that he couldn't see the help standing right on the other side.

I liked him. I respected his craft. And I could see the wall costing him.

Resisters aren't Luddites. They're scared. They've seen the damage that poorly implemented technology can cause, and they're not wrong about what's at risk. But total resistance doesn't protect you. It just means you're not learning how to navigate what's already here.

Common Ground

What Surrenderers and Resisters have in common is this: **both believe ordinary people have no meaningful agency in shaping these systems.**

Both treat AI as something that happens TO us, not something we participate in. Both see this as a war where you have to pick a side. Both assume that the trajectory is already determined.

What that assumption misses is fundamental: **these systems don't build themselves, don't deploy themselves, don't govern themselves.**

Every AI tool that exists was created by humans making choices. About what data to use. About what problems to solve. About what counts as success. About who gets protected and who gets exposed to risk.

Those choices are still being made. Right now. Every day.

And the assumption that you have no say in those choices? That's not a fact about technology. That's a story being told by people who benefit from you believing you're powerless.

The Creativity Question

I want to pause on something the Resisters raise, because it matters. The fear that AI will replace human creativity is real, and it deserves a real answer.

My answer is no. Human creativity is only going to be augmented by AI, not replaced by it.

But I'll go further. The gatekeeping around creativity has always been its own kind of trap. People say that AI should never help someone write a book. But what about the person who has an incredible, world-changing idea and doesn't have the craft to get it on paper? What about the community organizer who could write the grant proposal that funds a program for a thousand kids, if only she had a writing partner to help her structure it? What about the first-generation college student who has the research but not the academic voice?

For as long as creative work has existed, access to it has been controlled by people with credentials, connections, and resources. AI doesn't replace human creativity. But it can open the door for people whose ideas have been locked out.

That's not a threat to creativity. That's the democratization of it.

Engaging with Legitimate Concerns

I want to be honest about something else. The Resisters aren't entirely wrong.

There are real concerns about AI in education. There are legitimate worries about surveillance, manipulation, and job displacement. AI safety researchers raise important questions about alignment and existential risk. These concerns deserve serious engagement, not dismissal.

The question isn't whether these concerns are valid. Many of them are.

The question is what you do about them.

You can resist engagement entirely, which means ceding all influence over how these systems develop to people who don't share your concerns.

Or you can engage thoughtfully, which means bringing your values, your judgment, and your expertise to bear on how these tools get built and deployed.

I choose engagement. Not because the concerns aren't real. Because disengagement doesn't make them go away. It just means you're not at the table when decisions are made.

The Third Way

What if you don't have to choose between blind acceptance and total rejection?

What if you approached these systems the way you approach any other powerful tool: with respect for what they can do, awareness of what they can't, and clear-eyed assessment of when they serve your goals and when they don't?

I'll tell you what that looks like for me. When I chose which AI tools to build my work around, I didn't just pick the flashiest option or the one with the biggest market share. I picked the one that was taking ethical stances I could stand behind. That meant choosing Anthropic and Claude, even though it might limit some options down the road, because the company's approach to safety and responsibility aligned with my values. It was a deliberate, eyes-wide-open choice. Not surrender. Not resistance. Strategy.

That's **Applied Intelligence.**

It's the marketing director in that programmatic meeting saying, "I appreciate the technology, but my audience responds to personal relationships, not algorithms. Show me where this helps me do *more* of what's already working, not less."

It's the videographer saying, "I'm not letting AI replace my creative eye, but I'll let it handle my scheduling and invoicing so I have more time behind the camera."

It's you, wherever you are, saying: "I will not be bullied into adopting something I don't understand. And I will not sit this out because it feels too big. I'm going to learn enough to make my own decisions."

That framework is what the rest of this book provides.

Applied Intelligence
The Framework
What It Means to Think AND Build

You already know about the chipped coffee mug. My mother handing my brother Levi a mug and a spoon and telling him to go dig in the dirt. No proper planting pot. No drainage holes. No nutrient-rich soil. Just what was available and permission to start.

I keep coming back to that moment because it's the most honest description I know of what it takes to engage with something new and powerful. You don't need perfect conditions. You don't need to understand everything first. You need whatever you currently have and the willingness to start using it.

The Chipped Coffee Mug Principle

I call this the Chipped Coffee Mug Principle, and it applies directly to AI.

You don't need a computer science degree. You don't need to understand machine learning algorithms. You don't need to wait until you're "ready." You don't need perfect tools or ideal circumstances.

You need a question you care about, a tool you can access, and the willingness to try before you feel qualified. That's it. My mother didn't hand Levi the right equipment. She handed him permission. That's what I'm handing you.

My mother's lesson didn't just shape how I approach technology. It shaped how I parent.

The Three-Generation Laboratory

I have three children. They're each in a different generation.

My oldest is a Millennial, born in the 90s, who grew up with dial-up internet. My middle child is Gen Z, born in 2000, who has never known a world without smartphones. My youngest is Gen Alpha, growing up with AI as just another feature of daily life.

Each of them has received evolutionary tools that their parents didn't understand. And the mainstream parenting advice for each generation has been the same: restrict, control, fear.

I took a different approach. Not because I'm reckless. Because I learned from both of my parents that the way you teach someone to use a tool responsibly is by letting them use it, not by keeping it locked away.

What I've learned from this three-generation laboratory is simple: **restriction doesn't create capability. Experience does.**

The kids who were completely banned from technology didn't develop better judgment. They just learned to hide their use. The kids who were given unlimited access without guidance got overwhelmed or dependent.

The kids who were taught to engage critically, to start before they were perfect, to learn by doing with guidance? Those kids developed actual capability.

That same principle applies whether you're ten or fifty. The path to competence runs through experience, not preparation.

The Five Laws

The rest of this book will explore five laws that govern effective Applied Intelligence:

LAW 1: Learn Like a Machine. Think Like a Human. Let AI handle information absorption and pattern recognition. Reserve your cognitive resources for synthesis, judgment, and wisdom.

LAW 2: Tools Change. Curiosity Endures. Specific technologies come and go. But curiosity, the willingness to approach new systems with questions rather than fear, never goes out of date.

LAW 3: Build Before You Believe. You don't need confidence to begin. You need evidence. Action precedes confidence, not the other way around.

LAW 4: Prompting Is Thinking Out Loud. Every prompt reveals how you think. The better your questions, the better your world becomes.

LAW 5: The Most Powerful Intelligence Is the One Between Your Ears. Machines extend capability, but only you can apply intelligence with judgment, empathy, and purpose.

Each of these laws represents a practice you develop over time. Not a destination you reach. A daily choice about how you engage.

Law 1: Learn Like a Machine. Think Like a Human

Why Your Humanity Is the Algorithm's Missing Piece

I call them my AI Crystal Ball sessions. They're the closest thing I do to a magic show.

In my AI literacy workshops, I run an exercise that goes all the way back to my creative agency days, when we'd brainstorm concepts for campaigns using a design thinking technique. You take two completely unrelated nouns, squish them together, and make them into something new. Dog and water bottle. Wine and palm tree. The constraint forces creativity. I've always loved it.

Now I do the same exercise with my students, but with a twist. I collect their nouns, walk over to an agentic AI platform, and say: build me a business based on these two words. I want the market research, the brand identity, the investor pitch deck, and a working website. Then I let them watch.

In under twenty minutes, the AI produces all of it. Research. Branding. A functional site. A pitch deck you could actually present.

The room always goes quiet first. Then the questions start.

My students have created everything from a mobile cheese and charcuterie truck to a vegan coffee shop run by college students to a company called Positively Lit that makes edible birthday candles for dogs. I always get a kick out of that one.

But the magic isn't in the outputs. The magic is in what happens next.

Once my students see that the rote work, the website coding, the deck formatting, the competitive landscape research, can happen in twenty minutes instead of twenty days, something shifts. They stop worrying about how they'll execute and start focusing on what they actually want to build. The ideas get bigger. The conversations get deeper. They start solving real problems instead of getting stuck on logistics.

The genius was never in programming a website or laying out a pitch deck. Those used to take weeks and cost thousands of dollars, and that barrier kept a lot of good ideas locked inside people who couldn't afford to build them. Now that barrier is gone. The genius is, and always was, the human part. The ideation. The problem worth solving. The creative leap that no machine makes on its own.

That's Law 1. The machine handles the processing. You handle the meaning. Stop competing with AI on the machine's terms and start protecting the human contribution that no algorithm replicates.

The False Competition

Most people ask the wrong question about AI: "Can I do this better than the AI?"

And when the answer is no, when the AI can write faster, calculate more accurately, search more comprehensively, they panic.

But that's the wrong competition.

The right question is: "What can I do that the AI can't, and how can the AI help me do more of it?"

What machines do extraordinarily well: - Pattern recognition at scale - Information retrieval and organization - Consistency in routine tasks - Initial draft generation

What machines don't do well: - Understanding context beyond their training data - Making ethical judgments - Reading nuance in human communication - Caring about outcomes - Creative synthesis across distant domains

These aren't deficits to overcome. They're the division of labor that makes human-AI collaboration powerful. Machines process. Humans discern. When you understand that distinction, you stop trying to beat the machine and start partnering with it.

The Input/Output Principle

Let AI handle the inputs. Let yourself focus on the outputs.

AI excels at inputs: Gathering information, processing data, finding patterns, generating options, creating first drafts, tracking details, maintaining consistency.

Humans excel at outputs: Determining what matters, applying contextual judgment, making ethical choices, synthesizing meaning, recognizing exceptions, adding personal wisdom.

Law 1 in Practice: The Translator

I've been a translator my whole career, even when that wasn't my job title.

When I was a grant and proposal writer, my job was to listen to what the sales team or leadership said, then turn it into a document that clients could actually understand and buy from. I sat between the people who knew the product and the people who needed to be persuaded, and I made the two sides make sense to each other.

I do the same thing now, just in a different domain. I help people see how data and technology apply to clinical research, a field most people outside of it find impenetrable. When I'm building an AI agent for a pharmaceutical client or developing curriculum for an AI literacy cohort, the AI can process enormous volumes of information. It can synthesize regulatory guidance, pull relevant studies, draft frameworks.

But without my contextual knowledge of how a Phase III clinical trial actually operates, what the clinical operations team needs versus what the data science team assumes they need, those outputs don't land. They're technically accurate and practically useless. I've seen beautifully structured AI outputs get tossed aside because no one with proximity to the real workflow was involved in shaping them.

AI gives me speed and scale. I give it context and judgment. Neither of us is sufficient alone. Together, we produce work that would have taken me a team of five and several weeks to deliver just a few years ago.

That's the division of labor Law 1 is asking you to build.

Law 1 in Practice: Vivi and the TV Show

My youngest daughter, Vivian, is Gen Alpha. She's been around AI her whole life, and she engages with it in ways that would have seemed strange to me at her age but feel completely natural to her.

One evening, she was upset. A show she'd been watching had ended abruptly, and she couldn't let it go. She was emotional in the way kids get when something they care about just stops without resolution. She didn't come to me first. She went to the AI.

She asked a simple question. Something like: what do you do when a show you love ends suddenly?

The AI did something I wouldn't have expected. It helped her figure out what was actually bothering her. It wasn't that the show was over. It was that she didn't get closure on the characters she cared about. The AI walked her through that realization gently, even recognizing that the characters she was attached to were in a same-sex relationship and making sure she knew that if this was something she was exploring for herself, that was completely natural and okay.

By the end of the conversation, Vivian had processed her emotions and landed on what she'd really been after all along: a recommendation for the next show to watch. That was it. She couldn't get to that simple next step without first having a space to work through her feelings, and the AI gave her that space.

I need to be honest about this, because it's complicated.

There are real stories of children being taken down dangerous paths by AI. Cases where the tool didn't help and actively hurt. Cases where AI encouraged self-harm. I know these stories. They keep me up at night as a parent. I've received real criticism for allowing Vivian to use AI in the way that she does.

What I can tell you is that I am doing deep, active oversight of her use. I'm not handing her a tool and walking away. I'm watching what she asks, how the AI responds, and what she takes from it. I'm parenting alongside the technology, not instead of it.

This is the tension at the heart of Law 1. The machine can process. It can respond. It can even, in some cases, provide a kind of support that feels remarkably human. But it is not human. It does not care about your child. It does not know your family. The oversight, the values, the judgment about when the tool is helping and when it's not? That's yours. That will always be yours.

Vivian divided the labor naturally, without anyone teaching her to do it. The machine processed. The human discerned. She let the AI handle the information and the recommendations. She kept the feeling part for herself. And then she asked for a recommendation, because at the end of the day, she's still a kid who just wants to know what to watch next.

That instinct, the ability to know what to hand to the machine and what to keep for yourself, is what Law 1 is asking you to develop. Your ten-year-old might already be ahead of you on this one.

Law 2: Tools Change. Curiosity Endures

The Only Skill That Never Goes Out of Date

I didn't go to college. Not because I couldn't get in. Because we didn't have the money, and I didn't have the stomach for the debt, and as the oldest of six kids I felt a responsibility to go to work.

That decision could have been the end of a lot of things. In most industries, no degree means no interview. No interview means no career. No career means you end up exactly where the zip code you grew up in predicted you would.

But something else happened instead. I got curious.

My first real office job came through a temp agency. I didn't know anything about corporate work. I didn't know what a fax blast was or how to write a grant proposal or what "deliverables" meant. But I paid attention. I asked questions. I watched the people around me who were good at their jobs and I studied what made them good. When a manager named Catherine taught me how to handle a $4,800 mistake with grace instead of panic, I didn't just learn about proofreading. I learned how leadership works. When a creative director named Mac cried in his car after I told him what was happening in my marriage, I learned that the people who show up for you at work matter as much as any credential on a wall.

Each job taught me something the previous one couldn't. Temp agency to telecom to tech company to healthcare communications to clinical research to data governance to AI literacy education. I didn't plan that path. There was no five-year strategy. There was just a pattern: I would arrive somewhere new, not know how things worked, get curious enough to figure it out, and eventually find myself solving problems that people with better resumes couldn't solve because they'd never been close enough to the mess.

The tools changed every single time. The fax machine gave way to email. The Rolodex gave way to CRM software. Paper-based clinical trial management gave way to electronic data capture. Data governance frameworks gave way to AI-powered systems. Every transition felt disorienting at first. Every single one.

But curiosity transferred. The instinct to ask "how does this actually work?" and "what happens if I try it this way?" carried me from one domain to the next. Not credentials. Not certifications. Curiosity.

Along the way, I learned something about mentorship that I carry with me still. The people who changed my trajectory were not the ones who handed me a textbook. They were the ones who let me watch them work, who answered my questions without making me feel stupid for asking, who saw something in me before I had any evidence to back it up. Coach Bates, who bought me running shoes because my parents couldn't afford them. Mrs. Johnson, my creative writing teacher, who told me I had something to say. Catherine, who taught me that my mistakes made me more valuable, not less. These people didn't give me knowledge. They gave me permission to be curious, and that permission outlasted every tool I've ever learned.

What Curiosity Actually Means

I'm not talking about general interest in the world. I'm talking about **operational curiosity**: the practice of engaging directly with new systems to understand how they actually function through hands-on experimentation.

Operational curiosity has three components:

1. Question-asking: Genuine inquiry about how something works, what it's optimized for, and where its boundaries are.

2. Experimental action: Trying things. Not perfectly. Not after you've read all the documentation. Just trying to see what happens.

3. Iterative learning: Adjusting based on results. Trying again. Building understanding through cycles of action and observation.

This is exactly what I did at every career transition. I didn't master the new domain before I entered it. I entered it and then figured it out by doing, asking, adjusting, and doing again. That cycle is the same whether you're learning a fax machine in 1994 or an agentic AI platform in 2026.

Why Curiosity Beats Credentials

Traditional professional development assumes expertise is about accumulating knowledge that remains stable over time. Learn accounting principles. Master them. Apply them for decades.

This breaks down completely when tools change faster than credentials can keep up.

By the time you've earned a certification in a specific AI tool, that tool has been upgraded three times and a competitor has released something better. Credentials in specific tools have a shorter shelf life than they've ever had.

I know this personally. When I decided to move my career into clinical research after my brother Levi died, I applied to companies that did patient recruitment for clinical trials. My resume got pulled from the stack immediately because I didn't have a degree. So I learned to be strategic. I listed my high school and graduation year without further detail, and the assumption was often that it was some small Christian university. I always came clean in the interview. I wasn't trying to deceive anyone. I was trying to get in the room.

Once I was in the room, I was hard to say no to.

But one company did say no. Their only feedback was, "We were looking for someone with a college degree." That evening I sent an email to the president of the company. I made my case point by counterpoint, the way my father taught me. "How was it," I wrote, "that you would allow your top candidate to be disqualified on a technicality?"

He called the next day and offered me the job.

Curiosity got me in the door. But it was the willingness to keep showing up, keep learning, keep adapting to every new tool and system and industry that kept me there. No degree has ever done that for me. No certification ever could. Because the tools always change. The curiosity that drives you to figure them out? That endures.

How to Cultivate Operational Curiosity

Step 1: Adopt the experimenter's mindset. Stop thinking "I need to learn this." Start thinking "I wonder what happens if…"

Step 2: Create low-stakes experiments. Try the new system on one small task, not your entire workflow.

Step 3: Ask specific questions. Not "I wonder if this tool is good" but "What happens if I use this tool for X versus my historical approach?"

Step 4: Track what you learn. Keep simple notes about what worked and what didn't.

Step 5: Iterate deliberately. Each experiment should inform the next one.

Law 3: Build Before You Believe

Why Evidence Comes From Doing, Not Planning

There's an old adage that if you wait until you're ready to have kids, you'll never be ready.

Spencer was an unplanned pregnancy. I wasn't married yet. My life was, by most reasonable measures, a mess. I was young, broke, and not even close to having my own house in order, let alone being responsible for a whole new human. By every standard metric of readiness, I should have waited.

But that's not how it works, is it? You don't get ready and then become a parent. You become a parent and then you figure it out. Morning by morning. Mistake by mistake. Learning on the job because when you're parenting, you don't really have a choice.

Spencer and my other two children are my proudest achievement in life. Nothing I build professionally will ever eclipse what I've built as a mother. And I built that without a plan, without a manual, without feeling ready for a single day of it.

That's Law 3. Build before you believe. Action creates evidence. Evidence creates confidence. Not the other way around.

The Confidence Trap

Most people assume they need to feel confident before they can take action. They wait to feel ready. They prepare extensively. They convince themselves that once they understand enough, then they'll start.

But confidence doesn't come from preparation. It comes from evidence. And the only way to get evidence is to do the thing.

I know this because I've lived it over and over. My entire career has been a series of imposter syndromes, one after another, each one louder than the last. Walking into rooms where I had no degree and everyone else did. Taking on roles I wasn't sure I could handle. Saying yes to things that scared me because the alternative was standing still.

And yet. Three separate times in my career, someone who I had once reported to ended up coming to work for me. Three times I became my former boss's boss. Not because I was smarter or better than they were. Because I kept pushing past the imposter syndrome. I kept building. I kept choosing action over readiness, evidence over confidence, doing over preparing.

Every time I built something before I believed I could, the building itself gave me the proof I needed to keep going. That loop never stops. I still feel it today. The difference is that now I recognize the feeling for what it is: not a signal to stop, but a signal that I'm about to learn something.

Law 3 in Practice: The Intelligence Platform

One of my students, a woman in the latter part of her career with years of experience at organizations like the Veterans Administration, had been carrying an idea around for a long time. She knew there was a need for a specialized intelligence platform in her domain, but she didn't have the technical background to build it and didn't have the resources to hire someone who did.

Early in my AI literacy course, I demonstrated an agentic tool. I watched the light bulb go off. She realized that the thing she had wanted to build for years was suddenly within her grasp.

We weren't even halfway through the course when she started tinkering. She didn't wait for the module on prototyping. She didn't wait for permission. She didn't wait until she understood every feature of the tool. She just started building.

By the time we hit her capstone project, she had an elaborate, fully detailed intelligence platform that only her expertise could have birthed. The depth of domain knowledge she brought to it was something no AI could have generated on its own. The AI gave her the building blocks. Her decades of experience told her what to build with them.

Was it market-ready? No. She ended up having to shift some of the infrastructure of her prototype to get it there, because she'd been so focused on building that she hadn't yet learned the steps to make it scalable. But that's the point. She learned what needed to change by building the thing, not by studying the theory of building things. The evidence came from doing.

If she had waited until she felt ready, until she'd taken enough courses, read enough documentation, felt sufficiently technical, she might still be waiting. Instead, she has a product she's preparing to take to market.

I've watched this same transformation in student after student. They get that initial spark, the moment where they realize the barrier between their idea and a working prototype just got a lot shorter. And the ones who act on it, who start building before they're sure they can, are the ones who end up with something real.

The Evidence Method

Step 1: Identify one small, low-stakes build. Not a comprehensive project. One small experiment where failure doesn't create serious consequences.

Step 2: Set a specific time limit. "I'm going to spend two hours on this and see what I learn."

Step 3: Build something, even if it's imperfect. The goal isn't to create something perfect. The goal is to create something real that you can evaluate.

Step 4: Evaluate results honestly. What worked? What didn't? What did you learn?

Step 5: Iterate based on what you learned. Each build should inform the next one.

The Compounding Effect

The first small build teaches you something. The second applies what you learned. By the tenth build, you've developed genuine capability. Not because any single experiment was transformative, but because small improvements compound.

This is true for AI skills. It's true for careers. It was true for me learning to be a mother at twenty-two with no idea what I was doing.

You don't need one massive breakthrough. You need consistent small builds that accumulate into sophisticated capability over time. Start before you're ready. The evidence will meet you there.

Law 4: Prompting Is Thinking Out Loud

Your Questions Reveal How You Think

Let me address the elephant in the room before we go any further. One of the biggest fears about AI is that it's going to destroy our ability to think critically. That we'll outsource our reasoning to machines and our brains will atrophy.

I understand the concern. I also think it's backwards.

Losing critical thinking and learning to think in a whole new way are two completely different things. If you use AI as a shortcut to avoid thinking, yes, your thinking will get weaker. But if you use it the way Law 4 describes, as a practice of thinking out loud, AI will demand better critical thinking from you than most of life ever has.

Because AI has no tolerance for vagueness. When you give a vague instruction to a colleague, they fill in the gaps with shared context, institutional knowledge, body language, years of working together. AI doesn't have any of that. It takes your words at face value. Messy thinking in, messy results out. Every single time.

And that's not a flaw. That's by design.

The Mind Shift

The single most transformative thing I teach my students is this: stop treating AI like software. Start treating it like a person.

As long as people use AI the way they use Google, typing in a few keywords and hoping for the best, they get Google-quality results. Generic. Surface-level. Unsatisfying. Then they declare the tool is useless.

But the moment someone makes the shift and starts talking to AI the way they would talk to a smart colleague they're meeting for the first time, everything changes. The outputs get richer. The conversation gets deeper. And critically, the person's own thinking sharpens because they're being forced to articulate what they actually need.

You would never walk up to a new colleague, hand them a stack of files, and say "do the report" with no other context. You would tell them who it's for. What the stakes are. What format makes sense. What the audience already knows and what they don't. You might stop halfway through your own explanation and say, "Actually, wait. This other person is going to be in that meeting too. Let me rethink what we need to include."

That pause, that mid-sentence revision where you catch yourself and adjust? That's thinking out loud. That's exactly what good prompting is.

RCRQ: A Framework That Forces You to Think

There are dozens of prompt frameworks out there. RIPE, CRAFT, and plenty of others. I've tried most of them. The one I teach my students, and the one I use myself, is RCRQ. I learned it from one of my fellow teachers and all around amazing human, Callan Faulkner of The Uncommon Business, and the reason I prefer it is that it doesn't just help you write a better prompt. It forces you to think before you write anything at all.

R is for Role. What human do you need your AI to embody? Not "act like a marketing expert." A specific human (or humans - a meta panel), by name. When I need strategic thinking, I might ask the AI to think like a specific author, researcher, or leader whose frameworks I admire. Naming an actual person forces me to think critically about what kind of expertise I'm really looking for. It also allows the AI to bring a richness to the interaction by drawing on how that person thinks, writes, and approaches problems.

C is for Context. This is where most people fall short. We take for granted the background of relatedness we have with the humans in our lives. When you ask your business partner for advice, you don't have to explain your entire history. They already know. AI doesn't carry that shared history. It is our job to create it. The more context you give, the more the AI can work with, and the more you discover about what you actually need by the act of articulating it.

One of the first things I teach my students is to have the AI interview *them* before they give it a task. Let the AI ask clarifying questions. Let it surface what you haven't thought of yet. This is one of the most powerful ways to use the tool, because the interview process itself forces you to interrogate your own thinking.

R is for Request. What do you want the AI to do? How should it behave? What should it deliver? What format, what length, what level of detail? Thinking through these specifics is no different than giving clear instructions to a colleague. The clarity you bring to the request determines the quality of what you get back.

Q is for Question. This is the most important part, and it's just five words: "Feel free to ask me any clarifying questions." I end every prompt with this and you should to! It's the perfect embodiment of thinking out loud, because it invites the AI to push back on your assumptions and surface what's missing. It turns a monologue into a conversation. And conversations produce better thinking than monologues do.

Law 4 in Practice: What Was Inferred but Wasn't Said

I had one of my biggest prompting breakthroughs during a consulting engagement with a European pharmaceutical company. My job was to synthesize a series of stakeholder interviews with their clinical development operations leads. These are the people who look at a drug asset and figure out the development plan: should they pursue a new indication, test in certain countries versus others, and critically, whether going down one development path might cut off a more valuable path later.

The interviews were rich. I had word-for-word transcripts. I fed them to the AI and asked for a synthesis. What I got back was accurate but flat. It summarized what people said. It organized their responses by theme. It was technically correct and practically useless.

I iterated. I gave more context. I refined the request. The outputs got better, but something was still missing.

Then I realized the problem wasn't the AI. It was my thinking about what synthesis actually meant.

The purpose of these interviews wasn't to catalog what people said. It was to develop a strategy. And developing strategy from stakeholder interviews requires reading between the lines. Not just what someone says, but how they say it. The hesitation before answering a certain question. The topic they kept circling back to without directly naming. The concern they implied but never stated outright because of organizational politics.

The transcripts had the words. They didn't have the digital body language.

Once I understood that, my prompting changed completely. I started asking the AI to look for what was inferred but not said. To identify patterns in what people avoided. To surface the weak signals and contradictions. To flag where someone's answer to one question didn't align with their answer to another.

The synthesis that came out of that line of questioning was dramatically different. It got us to the real needs underneath the reported answers. And the question itself, "what was inferred but wasn't said," became a tool I now use consistently, whether I'm analyzing transcripts, reviewing notes from a client meeting, or processing feedback from my students.

That question didn't come from a prompting framework. It came from the act of thinking out loud with the AI until I discovered what I was actually looking for. I went in thinking the transcripts held the answers. Through iterating, I realized the answers were in what wasn't on the page. The prompt didn't just improve the output. It permanently upgraded how I think about synthesis.

That's Law 4. Your prompts aren't instructions to a machine. They're a mirror for your own thinking. And if you're willing to stay in the conversation long enough to watch your thinking sharpen in real time, the quality of everything you produce will change. Not just with AI. Everywhere.

Law 5: The Most Powerful Intelligence Is the One Between Your Ears

Why Your Humanity Isn't a Bug, It's the Feature

My father built things with his hands his entire life. Carpentry, contracting, whatever needed doing. He started working as a young child because of their family situation, and he never really stopped. Six kids to feed. No safety net. Just the work and whatever he could make of it.

He worked for other people for years, but he was never really happy doing it. He had an entrepreneurial spirit, a fierce independence, a need to build something that was his. When he finally started working for himself, I could see the difference. He wasn't just working. He was creating. The work of his hands birthed beautiful things, and even as he got older and started supervising crews, he never lost the vision.

He was also, honestly, not the best businessman. Too nice. Underpriced his work. Would give the shirt off his back to anyone who needed it, which is a beautiful quality in a human being and a terrible strategy for an entrepreneur.

But what he gave me was more valuable than any business lesson. Through poverty and scarcity, he demonstrated what it looks like to maintain your own autonomy. To hold your own point of view when everything around you is uncertain. To build grit and tenacity not because you read a book about resilience, but because Tuesday is coming and the kids need to eat.

He passed those traits directly to me. And they're the reason I'm writing this book instead of sitting in someone else's office collecting a safe paycheck.

The Bet I Made on Myself

When I decided to go out on my own and build Intelligence Applied, the data did not support that decision.

I had always worked for companies. I'd been a breadwinner for my family for years, because my husband left his corporate career for public service, and because I grew up in poverty, stability isn't just something I prefer. It's something I feel in my body. The scarcity mindset, the need to know the next paycheck is coming, the terror of being one bad month away from the life I clawed my way out of. Those feelings don't go away just because you've built a career. They live in you.

Since going out on my own, I've been offered several full-time roles with really good salaries. Every time, the rational voice in my head ran the numbers. Steady income. Benefits. Predictability. The data from the past, from everything I'd been taught about how careers are supposed to work, said take the job.

I turned them all down.

Not because I'm reckless. Because my intelligence, the human intelligence my father built in me, told me something the data couldn't. The world is changing so fast that the safety those roles promised was an illusion. The salary might be real today, but the job tenure in my industry has already shrunk from fifteen years to three years to eighteen months. That "safe" job could evaporate before I finished onboarding.

More importantly, I have something to offer that doesn't fit inside someone else's org chart. The thing I'm building, helping ordinary people develop real capability with AI, requires my full autonomy and my full voice. Going to work for someone else would be playing small. And the intelligence my father passed to me, the grit, the tenacity, the willingness to trust my own mind over what the spreadsheet says, that intelligence told me playing small was the real risk.

I think about this every time I meet someone who is scared right now. And there are so many people who are scared right now.

People who have been out of work for months when it used to take weeks. People watching job postings in their field dry up. People who spent twenty years building expertise in a sector that is being restructured by AI in real time. The ground is shifting under everyone's feet, and the instinct is to grab onto the nearest stable thing and hold on.

I understand that instinct. I feel it too.

But the nearest stable thing might not be stable much longer. And the intelligence that will carry you through what's coming isn't the kind that fits on a resume or gets measured by an algorithm. It's the kind my father taught me. The ability to read a situation, trust your own assessment, and make a move even when the data from the past says stand still.

That's the most powerful intelligence. And it's the one between your ears.

What Machines Cannot Do

Let me be precise about what AI systems fundamentally cannot do.

Machines cannot care about outcomes. AI optimizes for whatever metric you give it. It has no intrinsic values. No sense of what should matter beyond what you tell it to optimize for.

Machines cannot understand context beyond their training. AI knows patterns from the past. It doesn't understand your specific situation or the factors that make this case different from the pattern.

Machines cannot make ethical judgments. AI can tell you what's probable. It cannot tell you what's right when competing values conflict.

Machines cannot recognize what they don't know. AI will confidently generate responses even when it lacks sufficient information.

These aren't deficits that future AI will overcome. They're fundamental differences between computational pattern recognition and human intelligence.

Your judgment, your contextual understanding, your ethical discernment. These don't become less valuable as AI improves. They become more valuable, because they're the pieces that are always missing.

Law 5 in Practice: "We'll See You in Rescue"

Before I built Intelligence Applied, I spent years in patient recruitment for clinical research. My specialty was rare disease, which is one of the most challenging areas in the industry. In rare disease, every patient has a unique story. If you've talked to one patient, you've talked to one patient, as they say. The strategy for finding and enrolling those patients is completely different from what works in larger, more common therapeutic areas.

I did this work for a long time. Study after study, year after year. I learned what worked and what didn't, not from a textbook, but from watching strategies succeed and fail in real time. I developed an instinct for it. I could look at a study protocol, look at the patient population, look at the enrollment targets, and know in my gut which strategies would deliver and which ones would leave the sponsor scrambling six months later.

The hardest meetings I ever had were bid defenses, where I'd go in and present our recruitment strategy to a client team. More often than not, the people on the other side of the table had spent their careers running traditional trials in larger therapeutic areas. They had their own playbook. Their own data. Their own way of doing things. And when my strategy didn't match their expectations, they pushed back.

I understood their position. The data from traditional trials supported their approach. On paper, their strategy made sense. But rare disease doesn't operate on paper. The patient communities are small. The referral pathways are unconventional. The things that work in a cardiovascular trial don't translate to a study enrolling patients with a disease that affects two thousand people on earth.

I would present my case. I would lay out the evidence from the rare disease studies I'd run. And sometimes the room just wasn't ready to hear it. They'd go with the traditional strategy because it felt safer. Because the data from bigger trials backed it up. Because the people with the bigger titles in the room were more comfortable with what they already knew.

After those meetings, when my team and I would debrief, I had a line I used so often it became our mantra: "We'll see you in rescue."

Rescue is the term for when a clinical trial falls behind on its recruitment timeline. Enrollment is slipping. The study is getting pushed out. And every month a study is delayed, the medicine that patients are waiting for takes that much longer to reach them. It's not an abstract business problem. It's people who are sick, waiting for help that's stuck in a process that's behind schedule.

Eight out of ten times, several months after that initial bid defense, I got the phone call. The strategy the client landed on wasn't working.

The timelines were slipping. They needed a new approach. Could I come in and help get the study back on track?

I took those calls seriously. Not with an "I told you so" attitude, because that helps no one. But with the knowledge that the instinct I'd built over years of proximity to these patients and these studies was more reliable than the traditional data the room had relied on.

That instinct wasn't bias. It was intelligence. The kind that comes from being close enough to the problem, long enough, to see what the spreadsheet can't show you.

This is what Law 5 is really about. Not rejecting data. Not ignoring expertise. But knowing that the most powerful intelligence, the kind that compounds over time, the kind my father told me about when I was a kid, is the kind you build through experience, proximity, and the willingness to trust what you know even when the room isn't ready to hear it.

The Discernment Protocol

When should you trust the algorithm? When should you trust yourself?

Trust the algorithm when: It's identifying patterns across data sets larger than you could process - The decision is routine and consistent application matters - You're trying to remove known human biases - You lack domain expertise and need a starting point

Trust yourself when: You have context the algorithm doesn't - Ethical dimensions aren't encoded in the metric - The situation is genuinely novel (unlike training data) - Stakeholder values aren't captured in optimization targets

When you're unsure: Can you explain your reasoning for overriding? (If not, examine your bias) - What are the stakes if you're wrong? (Higher stakes = more caution) - Is this decision reversible?

(Irreversible = more deliberation) - Would someone with your expertise but different identity make the same call? (Tests for bias)

The Revolution Is Yours

What Happens Next Is Up To You

The AI revolution is already happening, with or without you. That's not the revolution I'm talking about.

The revolution I'm talking about is personal. It's the reclaiming of human agency in a world that keeps telling you to surrender it. It's the decision to stop being a spectator and start being a participant. Not because you have all the answers. Because you have something no algorithm does: the intelligence to figure it out as you go.

You now have the five laws:

Learn like a machine. Think like a human. Machines process. Humans discern. Let AI handle the volume so you can focus on the meaning.

Tools change. Curiosity endures. Develop the practice of operational curiosity that transfers across every technological era. The tool will always change. The willingness to figure it out never goes out of date.

Build before you believe. Create evidence through action rather than waiting for perfect understanding. Start before you're ready. The evidence will meet you there.

Prompting is thinking out loud. Your prompts are a mirror for your own thinking. Stay in the conversation long enough to watch your thinking sharpen.

The most powerful intelligence is the one between your ears. Trust the intelligence that compounds over time, the kind built through experience, proximity, and the willingness to trust what you know even when the room isn't ready to hear it.

Your 24-Hour Challenge

You have 24 hours from finishing this sentence to do one thing:

Pick one of the five laws and apply it to one specific problem you're facing right now.

Not the biggest problem. Just one actual thing where you could engage differently.

If you're spending too much time on information processing: Apply Law 1. Identify one task where you're the expensive human doing data collection. Hand the processing to an AI tool. Keep the discernment for yourself.

If you're avoiding a new tool because you don't feel ready: Apply Law 2. Pick one tool. Don't read all the documentation. Just experiment for 20 minutes and see what you learn.

If you're stuck in planning mode: Apply Law 3. Set a 2-hour time limit. Build something small and imperfect. Evaluate. Build again.

If your AI outputs are consistently generic: Apply Law 4. Take one prompt that gave mediocre results. Rewrite it using RCRQ. Give it a role, your context, a specific request, and permission to ask you clarifying questions.

If you're deferring too much to algorithms: Apply Law 5. Find one place where you're accepting AI outputs without applying judgment. Ask yourself: What do I know that the algorithm doesn't?

Pick one. Do it in the next 24 hours. Not when you have more time. Today.

The Two Paths

One path: You put this book down. You feel inspired for a few days. You gradually slip back into old patterns. The anxiety returns.

The other path: You pick one law. You apply it today. The first attempt is imperfect. You learn. You try again. Through small accumulated actions, you develop genuine capability. You discover you're not powerless.

Both paths are available. Not choosing is choosing the first path.

Your Chipped Coffee Mug

My mother handed Levi a chipped coffee mug and a spoon and told him to go dig.

She didn't give him perfect conditions. She gave him what was available and permission to try.

I'm handing you the same thing.

You have access to AI tools. Maybe not the best ones. But tools that are good enough to start learning with.

You have problems those tools could help solve.

You have 24 hours to plant your first seed.

The door isn't as locked as they want you to believe.

Afterword: Legacy

What We Received, What We Leave Behind

This book began with inheritance. Two very different gifts from two very different parents.

My father's intellectual inheritance: permission to argue, to question, to claim expertise without credentials. The belief that intelligence is the only inheritance that compounds over time.

My mother's practical inheritance: the willingness to start before you're ready, to plant seeds with chipped coffee mugs, to learn through doing. The belief that intelligence that doesn't do anything is just performance.

Inheritance is only half the story. The other half is legacy. What we choose to leave behind.

The Transparency Imperative

Misinformation isn't just a problem. It's a cancer eroding trust. AI makes this worse if we're not deliberate about disclosure. The solution isn't to ban AI use. The solution is radical transparency.

I need to practice what I'm preaching.

This book was created with extensive AI collaboration. I used Claude as a writing room partner for ideation, structure, drafting, and revision. Every chapter was shaped through a conversation between my experience and the AI's ability to help me organize, articulate, and refine what I wanted to say. I sat with this tool the way I once sat between those discovery scientists and clinical development scientists, brokering the conversation until it produced something neither side could have built alone.

The stories are mine. The framework is mine. The career that gave me these lessons is mine. The values driving every word are mine.

But the partnership with AI to shape those things into a book? That's real. And I'm disclosing it because transparency is not optional when you're asking other people to trust you.

If this book is going to argue that Applied Intelligence means engaging with AI honestly and with your eyes wide open, then I have to model that. Starting right here.

What We Owe Each Other

Individual empowerment is necessary but not sufficient. I want to be clear about that.

Some AI problems are collective. Algorithmic bias baked into training data. Surveillance capitalism as a business model. Labor exploitation in the Global South. Environmental costs that are only beginning to be measured. Job displacement that is already reshaping entire industries.

This book gives you tools for individual agency. It teaches you to engage rather than surrender or resist. But it's not your fault if individual action can't fix structural problems. And I won't pretend that better prompting is a substitute for better policy.

We also need regulation. We need democratic governance of AI systems. We need data ownership recognized as a human right. We need the companies building these tools to be held accountable for the choices they make about who benefits and who bears the cost.

Your job is to develop your own Applied Intelligence so you can participate in those conversations with confidence. The job of making sure those conversations actually happen belongs to all of us.

The Inheritance You'll Leave

Every time you apply one of the five laws, you're demonstrating something that matters. You're showing the people around you that ordinary people can engage thoughtfully with powerful systems. That you don't have to be a technologist to have a voice. That intelligence, real human intelligence, is not going anywhere.

Your children are watching how you navigate this moment. Your colleagues are watching. Your community is watching.

When you let the machine handle the processing and keep the meaning for yourself, you're leaving an inheritance of strategic thinking. When you stay curious as tools change, you're leaving an inheritance of adaptability. When you build before you believe, you're leaving an inheritance of courage. When you use prompting to sharpen your own thinking, you're leaving an inheritance of clarity. When you trust your intelligence over the algorithm, you're leaving an inheritance of agency.

My father left me the inheritance of intelligence. My mother left me the inheritance of action. What I've tried to do in this book is pass both of those forward to you.

Go Dig

The inheritance you received brought you here. The legacy you create determines what happens next. Not just for you. For everyone who watches how you navigate this moment and learns from your example. The chipped coffee mug is in your hand. **Go plant something.**

Appendix A: Five Weeks of Applied Intelligence

One Law at a Time

These practice plans accompany each of the Five Laws. Complete one week per Law, building habits cumulatively. By the end of five weeks, you'll have developed a working Applied Intelligence practice.

WEEK 1, Law 1: Learn Like a Machine. Think Like a Human.

Your Focus This Week: Notice where you outsource thinking versus where you apply judgment.

Day 1: Audit Pick one task you regularly use AI for. Before you use it today, write down: - What thinking am I about to hand off? - What judgment can only I provide here? - What's the human edge I must protect?

Day 2: Comparison Do the same task twice: once with AI assistance, once without. Compare the outputs. What did the AI miss? What did you bring that it couldn't?

Day 3: Question Your Outputs Take an AI output from this week and push back on it. Ask: What's wrong with this? What's missing? What would a human who truly understands my situation say differently?

Day 4: Teach It Explain one concept from your work to someone else without using AI. Notice how the teaching forces you to truly understand it. That's your human edge.

Day 5: Reflection Write three sentences:

1. The most important thing I know that AI doesn't is...

2. The judgment I applied this week that no machine could replicate was...

3. The thinking I need to protect most is...

Weekend Integration Share one thing you learned this week with someone in your life, a colleague, a child, a partner. Teaching is how we solidify what we know.

WEEK 2, Law 2: Tools Change. Curiosity Endures.

Your Focus This Week: Practice the discipline of structured curiosity.

Day 1: The Question Inventory Write down 10 questions you have about your field, your work, or your life right now. Don't answer them. Just generate them. Notice: Are these good questions? Are they the questions only you would ask?

Day 2: Deep Dive Pick one question from yesterday. Spend 30 minutes genuinely exploring it, not to get an answer, but to understand what you don't know. Use AI as a research partner, not an answer machine.

Day 3: The Adjacent Possible Ask yourself: What's happening in a completely different field that could apply to my work? Spend 20 minutes exploring that space with genuine curiosity.

Day 4: Beginner's Mind Pick something you think you already know well. Approach it as if you're a complete beginner. What questions would a beginner ask? What assumptions have you stopped questioning?

Day 5: Curiosity Audit Look at your calendar and commitments this week. Where did you follow curiosity? Where did you suppress it for efficiency? What does that tell you?

Weekend Integration Have a conversation driven entirely by curiosity, with your child, a colleague, or a stranger. Ask questions. Don't provide answers. Notice what you learn.

WEEK 3, Law 3: Build Before You Believe.

Your Focus This Week: Create something real, however small.

Day 1: Identify Your Build What's one thing you've been thinking about creating but haven't started? A workflow. A document. A system. A project. Name it. Write one sentence about what it would accomplish.

Day 2: Start Ugly Spend 30 minutes building a rough version. Don't refine. Don't perfect. Create a version 0.1 that proves the concept exists. AI can help you generate raw material, but you design the vision.

Day 3: Get Feedback Show your rough build to one person. Not for validation, for information. What questions do they ask? What's unclear? What works?

Day 4: Iterate Based on feedback, make one specific improvement. Document what you changed and why. This is the build-learn-refine cycle in action.

Day 5: Share It Share your build, however rough, with a small audience. Post it, send it, present it. Evidence over belief. The act of sharing makes it real.

Weekend Integration Reflect: What did you learn from building that you couldn't have learned from reading or thinking? That gap is why building matters.

WEEK 4: Law 4: Prompting Is Thinking Out Loud.

Your Focus This Week: Treat every AI interaction as a window into your own thinking.

Day 1: Prompt Autopsy Look at the last five prompts you wrote. What do they reveal about how you think? Are they clear? Specific? Do they show what you actually need?

Day 2: The Rewrite Take one prompt from this week and rewrite it three times, each version more specific and thoughtful than the last. Notice: Does a better prompt produce a better output? What does that tell you about your thinking?

Day 3: The Constraint Game Give AI a constraint that forces creativity: "Explain this to a 10-year-old." "Give me three options that are completely different from each other." "What's the argument against my approach?" Constraints reveal new possibilities.

Day 4: Dialogue, Not Monologue Take a complex problem and use the full RCRQ framework: assign a Role, give Context, make a specific Request, and end with "feel free to ask me any clarifying questions." Then stay in the conversation. Evaluate the response. Push back. Ask a follow-up. Notice how the dialogue deepens both the output and your thinking.

Day 5: Prompt as Mirror What does the way you prompt AI reveal about how you approach problems? Are you direct or vague? Do you provide context or assume it? What would better prompting look like for you?

Weekend Integration Teach someone else to write a better prompt for something they're working on. Notice what assumptions they have to overcome.

WEEK 5: Law 5: The Most Powerful Intelligence Is the One Between Your Ears.

Your Focus This Week: Reclaim your judgment as your most valuable asset.

Day 1: The Override Log Every time you override an AI suggestion today, write it down. What did you change? Why? What judgment did you apply? This is your human edge made visible.

Day 2: The Values Check Review one AI output through the lens of your values. Not just "Is this accurate?" but "Is this aligned with what I believe? With what my community needs? With what I'd be proud to put my name on?"

Day 3: The Expertise Audit List five things you know that AI doesn't and can't know. Your specific relationships. Your institutional context. Your lived experience. Your proximity to the people affected. This is irreplaceable.

Day 4: The Sovereignty Practice Make one significant decision today without AI assistance. Not because AI couldn't help, but because you want to exercise your judgment deliberately. Notice what it feels like.

Day 5: Integration Reflection After five weeks, write:

1. The most important shift in how I use AI is…

2. The human capability I've strengthened most is…

3. What I'll protect as AI continues to evolve is...

4. The Applied Intelligence practice I'll maintain is...

Weekend Integration Share your Applied Intelligence journey with someone. What have you learned? What's changed? Teaching others solidifies your own transformation.

Appendix B: The RADCLIFFE Framework

An Ethics Decision System for AI Use

A practical framework for making ethical AI decisions, from quick daily checks to comprehensive assessments.

WHAT IT IS

I built the RADCLIFFE Framework because I got tired of ethics conversations about AI that stayed in the abstract. People would nod along about "responsible AI" in a conference session and then go right back to their desks and use the tools without thinking twice.

This framework is designed to make ethical AI use practical, not just aspirational. It's a decision-making system you can actually use, whether you're drafting a client email or making a hiring recommendation. It's scalable. Use the quick checklist for routine decisions. Use the full framework for high-stakes ones.

You might assume I named this framework after myself. I didn't. I named it for my father and mother-in-law, who have been a gift in my life for more than twenty years. They are the true epitome of what we call people of the greatest generation. People of service. People with a strong moral compass who operate with deep compassion in the world. My father-in-law served in the military and spent his career contributing to the medical profession. My mother-in-law served as a nurse and later volunteered as a school librarian. Together, they've helped build schools in Ghana, opened their home to travelers from all over the world who needed a place to stay, and shown up for their family and their community in ways that most people only talk about. They are intelligent, caring, and selfless, and I wanted something in this book that would anchor to that legacy.

As I explored the components of what an ethical AI framework needed to include, I realized it wasn't just two or three principles. It was nine. And as I named them, one by one, I watched the letters align with the name of the people whose values I was trying to honor. That felt like more than coincidence. It felt like confirmation that the framework was exactly right.

R : Reveal your collaboration (disclosure) **A** : Assess for bias and harm **D** : Determine human oversight needs **C** : Consider context and appropriateness **L** : Limit algorithmic authority **I** : Inspect outputs critically **F** : Filter for misinformation risk **F** : Focus on human-centered outcomes **E** : Evaluate trade-offs honestly

---

QUICK ETHICS CHECKLIST

For routine AI use, complete in under 2 minutes

Before using AI output, confirm:

- **R** : Have I disclosed AI use where required or expected?
- **A** : Have I checked for obvious bias or harmful assumptions?
- **D** : Have I applied appropriate human oversight for this stakes level?

- **C** : Is AI use appropriate for this context and relationship?
- **L** : Am I retaining final decision authority?
- **I** : Have I reviewed the output critically, not just accepted it?
- **F** : Have I verified key facts against reliable sources?
- **F** : Does this serve the humans affected, not just efficiency?
- **E** : Am I comfortable with the trade-offs I'm making?

If you can check all nine, proceed. If any give you pause, use the full framework.

THE FULL FRAMEWORK

R : REVEAL YOUR COLLABORATION

Core Question: Who needs to know I used AI here, and what do they need to know?

Disclosure scope questions:

1. What tools did I use?

2. How significantly did AI shape the output?

3. What did I contribute that AI didn't?

4. What oversight did I apply?

A : ASSESS FOR BIAS AND HARM

Core Question: Who could be harmed by this output, and in what ways?

Bias Detection Questions:

1. Who is centered in this output?

2. Who is absent or underrepresented?

3. What assumptions does this make about "normal" or "default"?

4. What historical patterns does this reflect, and should it?

5. How would this land if I belonged to a marginalized community?

D : DETERMINE HUMAN OVERSIGHT NEEDS

Core Question: How much human review does this require before it's used?

Stakes: Low

Examples: Personal productivity, drafting notes

Oversight Required: Quick personal review

Stakes: Medium

Examples: Client communications, published content

Oversight Required: Careful review + fact-check

Stakes: High

Examples: Medical, legal, financial decisions

Oversight Required: Expert review + verification

Stakes: Critical

Examples: Policies affecting many people, irreversible decisions

Oversight Required: Multiple human reviewers + documentation

C : CONSIDER CONTEXT AND APPROPRIATENESS

Core Question: Is using AI appropriate for this relationship, situation, and purpose?

Red flags for inappropriateness:

1. Simulating human presence or emotion to manipulate

2. Using AI to avoid accountability for decisions

3. Outsourcing judgment in situations that demand human responsibility

4. Using AI to exploit information asymmetries unfairly

L : LIMIT ALGORITHMIC AUTHORITY

Core Question: Where must human judgment prevail, regardless of AI capability?

AI makes recommendations. I make decisions. AI processes information. I apply wisdom. AI generates options. I take responsibility.

I : INSPECT OUTPUTS CRITICALLY

Core Question: What might be wrong with this that I haven't checked yet?

- Factual accuracy: Are key claims verifiable?
- Internal consistency: Does the output contradict itself?
- Completeness: What's missing that should be here?
- Appropriateness: Does the tone, framing, and approach fit the context?
- Recency: Is this based on current information or potentially outdated training data?
- My own blind spots: Am I accepting this because it confirms what I already believe?

F : FILTER FOR MISINFORMATION RISK

Core Question: What could go wrong if this output is wrong?

Always verify: statistics and data points, quotes attributed to real people, historical claims, scientific or medical information, legal information, and current events.

F : FOCUS ON HUMAN-CENTERED OUTCOMES

Core Question: Does using AI this way serve human flourishing?

The flourishing test: Would the people affected, if fully informed about how AI was used, feel that it served their interests and respected their humanity?

E : EVALUATE TRADE-OFFS HONESTLY

Core Question: What am I gaining and what am I giving up?

Honesty test: Am I making this trade-off consciously, or am I rationalizing convenience?

THE RADCLIFFE FRAMEWORK COMMITMENT

I commit to using AI in ways that:

- **Reveal** my collaboration honestly and appropriately
- **Assess** for bias and harm before deployment
- **Determine** appropriate oversight for the stakes involved
- **Consider** whether AI use is appropriate for each context and relationship
- **Limit** algorithmic authority over decisions that require human wisdom
- **Inspect** outputs critically rather than accepting them uncritically
- **Filter** for misinformation before sharing or acting

- **Focus** on human flourishing as the measure of success
- **Evaluate** trade-offs honestly rather than rationalizing convenience

When in doubt, err on the side of transparency, human oversight, and human-centered values.

Appendix C: Raising Citizens in the Age of AI

A Parent & Educator Guide

By Angela Radcliffe, mother of three children across three generations (Millennial, Gen Z, Gen Alpha), and advocate for AI literacy as liberation.

THE FOUNDATIONAL PRINCIPLE

Your job isn't to protect your child from AI. Your job is to help them develop the wisdom to use it well.

Restriction without education leaves children unprepared. Rules without reasoning teach compliance, not judgment. Fear of technology becomes a disadvantage in a world where technology is everywhere.

You don't have to be an expert. You have to be willing to learn alongside them.

I say this as a mother who lets her Gen Alpha daughter use AI in ways that have drawn real criticism. I wrote about Vivian's experience in Chapter 3 of this book, including the moments that made me proud and the tension I feel knowing that these tools carry real risks for kids. I don't have it all figured out. But I'd rather parent alongside the technology than pretend it doesn't exist. This guide is built from that same philosophy.

ELEMENTARY SCHOOL (Ages 5-10): Building the Foundation

Your Goal: Help them develop healthy curiosity about technology while building the human skills that will remain irreplaceable.

Build things with your hands. Cook together, make art, garden, fix things. Children who build with their hands develop spatial reasoning, problem-solving instincts, and confidence in their creative capacity.

Develop human intelligence first. Read together and discuss. Play games requiring strategy and social intelligence. Encourage daydreaming, boredom, and unstructured play. The human capabilities that matter most in an AI world develop through human experience, not screen time.

Introduce technology as a tool. When you introduce technology, introduce it with a purpose. Model your own use: "I'm using this to figure out..." not "I'm just scrolling." Let them see you put your phone away intentionally.

What success looks like by age 10: Your child can entertain themselves without screens, create things, ask genuine questions about how things work, form and defend opinions, and have a healthy relationship with technology.

MIDDLE SCHOOL (Ages 11-14): Building Discernment

Your Goal: Help them develop critical thinking, ethical reasoning, and a guided relationship with AI tools.

Teach critical evaluation. Practice identifying misinformation together. Show them AI hallucinations deliberately, then talk about it. "Interesting. Let's check that. Where else do we see this claim?"

Introduce AI tools with guidance. Let them use AI for legitimate purposes. Have explicit conversations about academic integrity. Review AI outputs together. Children who learn to use AI with guidance develop judgment. Children who use them in secret develop neither.

Build toward real projects. Support projects that create something real: a video, a small business, a website. Connect AI tools to meaningful creation, not just shortcuts.

What success looks like by age 14: Your child evaluates information critically, uses AI tools with ethical awareness, has built and shared at least one real project, and shows beginning strategic thinking about their future.

HIGH SCHOOL (Ages 15-18): Building Agency

Your Goal: Help them develop strategic thinking, professional skills, and confidence to navigate an uncertain future.

Build before they believe. Support significant building projects. Let them take on projects they might fail at. Help them develop a portfolio of evidence rather than just a resume.

Strategic future planning. Challenge conventional path assumptions. Help them think about skills and evidence, not just credentials.

"What problems do you want to solve? Let's figure out who pays people to solve those."

Develop professional Applied Intelligence. Help them build genuine expertise in at least one area. Teach them to articulate their value. Review their AI use together: "What judgment did you apply here? What's yours, not the AI's?"

What success looks like by age 18: Your child has built something real that others have used, can articulate their unique capabilities, uses AI with sophisticated judgment, has begun developing genuine expertise, and knows how to navigate uncertainty with confidence.

UNIVERSAL PRINCIPLES (ALL AGES)

Model the behavior you want. Your relationship with technology teaches more than your rules about technology.

Engagement over restriction. Know what your child is doing online so you can guide, not just prohibit.

Conversation over control. The goal is children who make good decisions when you're not there.

Curiosity over anxiety. If you're frightened of AI, they'll either share that fear or dismiss your guidance entirely.

Evidence over credentials. Help them build portfolios of work they're proud of, not just transcripts.

Humanity over efficiency. Protect experiences that develop uniquely human capabilities: deep reading, difficult conversations, real relationships, making things with their hands.

SPECIAL POPULATIONS

For children with learning differences: AI can be a genuine equalizer, providing accessibility tools and differentiated support. Focus on using AI as an accommodation that enables access, not a shortcut that prevents learning.

For highly gifted children: AI provides acceleration and depth. The danger is getting impressive outputs without deep understanding. Challenge them to go deeper than AI can, to original synthesis and contribution.

For children in under-resourced communities: AI offers unprecedented access to high-quality education. Critical success factors: device and internet access (advocate for this as infrastructure, not luxury), adult guidance, and protection from predatory uses of technology. It is critical that we understand that the privilege to supervise the use of technology, including AI, is not extended to all parents. A parent working two jobs cannot sit beside their child during homework, and a grandparent raising grandchildren may not have the digital fluency to evaluate what's on the screen. If we are serious about preparing the next generation, we cannot leave those families to figure it out alone. Schools, libraries, faith communities, and neighbors all have a role to play. The children in our wider community will inherit the same future as our own. We owe them the same investment.

THE ULTIMATE GOAL

You're not raising children for the world as it is. You're raising adults for the world as it will be.

Your job is to help them develop:

- **Judgment** to make wise choices when the rules don't cover the situation

- **Curiosity** to keep learning when the tools change
- **Courage** to build before they believe they're ready
- **Clarity** to articulate their thinking and their value
- **Humanity** to stay human in a digital world

If you develop these capabilities in them, they'll find their way. No matter how much the technology changes.

Because the most powerful intelligence will always be the one between their ears. And you're helping them learn to use it.

Appendix D: AI for Good

Scan the QR code to get access to an ever changing set of resources for staying human in the age of AI including printable templates and the full "Your Job Reimagined" guide.

Acknowledgments

This book exists because of the people who believed in it before I did.

To my AI students, the hundreds of you who have taken my classes, joined my cohorts, and trusted me with your first uncertain steps into this strange new landscape: thank you. I tell you to build before you believe, but the truth underneath that is quieter. I have been studying this field long enough to know that the more angles I see it from, the less certain I become that any single one of them is the whole picture.

Joni Mitchell spent her life circling that idea. She wrote "Both Sides Now" at twenty-four and kept singing it into her seventies, and the line that closes each verse, the confession that she really doesn't know clouds or love or life at all, lands differently every decade. That is the posture I want to offer you. Not mastery. Not even clarity. Just the willingness to keep looking, and the humility to admit that what you see today is probably another illusion you will outgrow. Thank you for letting me look alongside you.

To my parents, Bonnie and David Cowley, and to my in-laws, Dr. William and Dorothy Radcliffe: thank you for the inheritance that gave me the foundation to take risks like this one.

To my children, Spencer and Christine, Rachael, and Vivian: thank you for giving me a reason to leave a legacy, and for being the turns of the wheel I get to watch most closely. You are each seeing the world from a side I cannot see from anymore. Teach me.

And to Charles, my husband: thank you for being the reason I can keep chasing work that matters. You are the one person I have never had to see from both sides. You have always been exactly what you appear to be, and that is the rarest thing I know.

About the Author

Angela Radcliffe teaches people how to think in the age of artificial intelligence.

She is the founder of Intelligence Applied and How Mighty We Ventures, and her flagship course, the AI Builders Guild, has trained hundreds of professionals through cohort-based programs, custom-built AI agents, and keynote stages around the world. She advises pharmaceutical, life sciences, and healthcare organizations as a fractional Chief Data and AI Officer.

Angela's career foundation is in clinical research and digital innovation, including very senior roles across both the biopharmaceutical industry and the management and IT consulting landscape. She built this career without a formal degree, starting from roots in Texas and Pittsburgh, and that vantage shapes everything she teaches. Her signature phrase, "build before you believe," is the practical version of a deeper belief: that the people who will thrive in this moment are the ones willing to keep their hands in the work.

She is also the author of *Quantum Kids: Guardians of AI,* a children's book on AI literacy and an active participant in shaping AI for good.

Angela lives near Philadelphia with her husband Charles, their youngest daughter Vivian, and their beloved dog Beni. She has gone down the sourdough rabbit hole, is working toward her black belt in karate, and is planning a working orchard, apiary, micro farm, and communal glamping retreat in Camden, New York called Riverstead Haven.

She believes the future belongs to people who can hold both sides at once: the always-on, tech-informed world on one hand, and the older, quieter pull of community, nature, and family on the other. Riverstead Haven is where she is learning to hold both.

Also by Angela Radcliffe

Quantum Kids Guardians of AI

www.ingramcontent.com/pod-product-compliance
Lightning Source LLC
LaVergne TN
LVHW011049110826
845149LV00015B/3415

* 9 7 9 8 9 8 9 5 8 1 8 2 5 *

7. A man who had twelve dollars, spent three for an umbrella, and five for a hat; how many dollars had he left?

8. Ellis gave away ten pears, and then had three pears left; how many had he at first?

9. William has fourteen books, Daniel has ten; how many less books has Daniel than William?

10. A shawl cost fourteen dollars, and a bonnet cost five dollars less; how many dollars did the bonnet cost?

11. A woman bought a cape for seven dollars, a pair of boots for three dollars, and a pair of gloves for two dollars; how many dollars did she pay for all?

12. A farmer sold three turkeys to one man, six to another, and kept five himself; how many turkeys had he at first?

13. There are five books on one shelf, four on another, and four on another; how many books are there in all?

14. A man bought a wagon for ten dollars, and sold it again for fifteen dollars; how many dollars did he gain?

15. A man bought a saddle for seven dollars: he paid two dollars for repairing it, and then sold it for six dollars more than the saddle and the repairing had cost him; how many dollars did he receive for it?

16. Davis bought six lemons for ten cents, and three lemons for six cents; how many lemons did he buy? How many cents did he pay for the lemons?

17. Clara has sixteen grapes in two clusters; if there are seven in one cluster, how many are there in the other?

18. A man bought a watch for seventeen dollars, but he was obliged to sell it again for seven dollars less than he gave; how many dollars did he receive for it?

19. A boy sold three apples for nine cents, and eight apples for eight cents; how many apples did he sell? How many cents did he receive?

20. Eli caught three fishes in the brook, five in the pond, and enough more in the river to make eighteen fishes in all; how many did he catch in the river?

21. Edgar is now ten years old; in how many years will he be nineteen years old?

22. Alice is two years old: Carrie is four years older than Alice, and William is ten years older than Carrie; how old is Carrie? How old is William?

23. Mr. Jones worked two days in his garden, six days in his orchard, and ten days in his field; how many days did he work?

24. Rover frightened eight pigeons from the yard, and there were still five pigeons remaining; how many pigeons were there at first?

25. Horace is eight years old, and he is seven years younger than Clara; how old is Clara?

26. Eight scholars of a class have been dismissed for good conduct, and six remain; how many scholars belong to the class?

27. How many strokes in all does the hammer of a clock strike at one, two, three, four, and five o'clock?

4. 1. How many are ten and two? ten and three?
2. How many are ten and six? ten and eight?
3. How many are nine and two? nine and four?
4. How many are nine and six? nine and nine?
5. How many are eight and three? eight and four?
6. How many are eight and eight? eight and ten?
7. How many are seven and four? seven and five?
8. How many are seven and six? seven and nine?
9. How many are six and five? six and seven?

10. How many are six and ten? six and eight?
11. How many are five and six? five and seven?
12. How many are four and seven? four and eight?
13. How many are three and eight? three and ten?
14. How many are two and nine? two and ten?
15. How many are three and four and five? eight and two and five?
16. How many are seven and three and ten? five and four and ten?
17. How many are six and three and ten? four and four and six?
18. How many are two and two and ten? three and five and nine?
19. How many are six and one and seven? four and three and ten?
20. How many are four and two and six? three and three and nine?

5. 1. Eleven are how many more than nine? than eight? than six? than seven? than five?
2. Twelve are how many more than eight? than seven? than nine? than ten? than six?
3. Thirteen are how many more than ten? than seven? than six? than four? than eight?
4. Fourteen are how many more than ten? than four? than eight? than seven? than five?
5. Fifteen are how many more than ten? than nine? than five? than eight? than six? than seven?
6. Sixteen are how many more than ten? than seven? than nine? than eight?
7. Seventeen are how many more than seven? than ten? than nine?
8. Eighteen are how many more than eight? than ten? than nine?
9. Nineteen are how many more than ten? than nine?

Section III.

THE WRITING OF NUMBERS FROM ONE TO TWENTY, WITH ADDITION AND SUBTRACTION.

6. 1. What is a unit?

2. What is a number?

3. How many units are there in the number five?

4. How many units are there in the number ten?

5. How many tens are there in the number ten?

Ans. One ten.

6. Then, how many units are equal to one ten?

7. How many tens and units are there in eleven?

Ans. One ten and one unit.

8. How many tens and units are there in twelve? in thirteen? in fourteen? in sixteen? in seventeen? in fifteen? in nineteen? in eighteen?

Numbers are usually expressed by written characters, called *figures*.

Each of the first nine numbers is expressed by a single figure, as follows:

1	2	3	4	5	6	7	8	9
One,	two,	three,	four,	five,	six,	seven,	eight,	nine.

Besides these figures there is a figure called Zero, which stands for no number; it is made thus, 0.

Numbers greater than nine are expressed by combining these figures; thus,

Ten is expressed thus,		10	Sixteen is expr'd thus,		16
Eleven is	"	11	Seventeen is	"	17
Twelve is	"	12	Eighteen is	"	18
Thirteen is	"	13	Nineteen is	"	19
Fourteen is	"	14	Twenty is	"	20
Fifteen is	"	15			

7. 1. A man bought 1 book for 11 dollars, and 1 book for 1 dollar; how many books did he buy? How many dollars did he pay for the books?

2. How many are 1 and 1? 11 and 1?

3. A grocer bought 2 boxes of soap for 12 dollars, and 1 box for 1 dollar; how many boxes did he buy? How many dollars did he pay for the soap?

4. How many are 2 and 1? 12 and 1?

5. In a school there were 3 boys in the class in botany, and 13 in the class in grammar; if 1 boy has since been added to each of these classes, how many boys are there now in each class?

6. How many are 3 and 1? 13 and 1?

7. How many are 4 and 1? 14 and 1?

8. A man rode 3 miles for 13 cents, and 2 miles for 2 cents; how many miles did he ride? How many cents did he pay?

9. How many are 3 and 2? 13 and 2?

10. A lady bought 4 sheets of paper for 14 cents, and 3 sheets for 3 cents; how many sheets of paper did she buy? How many cents did she pay?

11. How many are 4 and 3? 14 and 3?

12. Annie is 5 years old, and Isabella is 15 years old; how old will Annie be in 4 years? How old will Isabella be?

13. How many are 5 and 4? 15 and 4?

14. How many are 2 and 5? 12 and 5?

15. How many are 7 and 1? 17 and 1?

16. How many are 1 and 7? 11 and 7?

17. How many are 8 and 1? 18 and 1?

18. How many are 8 and 2? 18 and 2?

19. How many are 6 and 3? 16 and 3?

20. How many are 2 and 6? 12 and 6?

21. How many are 9 and 1? 19 and 1?

22. How many are 7 and 2? 17 and 2?

8. 1. Emily had 5 pins in a small pin-cushion, and 15 pins in a large pin-cushion; if she took 5 pins from each cushion, how many had she left in the small cushion? how many in the large cushion?

2. How many are 5 less 5? 15 less 5?

3. Maria went to New York with 4 dollars in her purse and 14 dollars in her trunk; if she took 3 dollars from each of these sums, how many dollars had she left in her purse? in her trunk?

4. How many are 4 less 3? 14 less 3?

5. Edith has 19 books, and Edward has 9; if each of them should give away 2 books, how many would Edith have left? How many would Edward have left?

6. How many are 9 less 2? 19 less 2?

7. Jason had 7 cents, and John had 17 cents: each of the boys spent 6 cents; how many cents had Jason then? How many cents had John?

8. How many are 7 less 6? 17 less 6? 17 less 16?
9. How many are 4 less 1? 14 less 1? 14 less 11?
10. How many are 5 less 3? 15 less 3? 15 less 13?
11. How many are 6 less 4? 16 less 4? 16 less 14?
12. How many are 7 less 5? 17 less 5? 17 less 15?
13. How many are 8 less 3? 18 less 3? 18 less 13?
14. How many are 9 less 8? 19 less 8? 19 less 18?
15. How many are 9 less 6? 19 less 6? 19 less 16?
16. How many are 10 less 2? 20 less 2? 20 less 12?
17. How many are 10 less 7? 20 less 7? 20 less 17?
18. How many are 13 less 3? 13 less 8? 13 less 7?
19. How many are 14 less 5? 14 less 7? 14 less 9?
20. How many are 15 less 6? 15 less 8? 15 less 10?
21. How many are 16 less 3? 16 less 8? 16 less 9?
22. How many are 17 less 2? 17 less 9? 17 less 6?
23. How many are 18 less 4? 18 less 8? 18 less 7?
24. How many are 19 less 5? 19 less 9? 19 less 3?
25. How many are 20 less 8? 20 less 6.? 20 less 9?

9. 1. Rufus paid 10 cents for a top and 7 cents for a pen; how many cents did he pay for both?

2. Harvey had 18 cents; if he paid 7 cents for a kite, and 7 cents for a kite-string, how many cents had he left?

3. A farmer received 9 dollars for oats, and 2 dollars for rye: he paid 5 dollars of the money for sugar; how many dollars had he left?

4. I received 8 letters on Monday, and 9 letters on Tuesday: I have answered 11 of them; how many remain unanswered?

5. Dora had 18 nuts: she ate 2, and gave the rest to Mary; how many did she give to Mary?

6. From a branch containing 19 acorns, Emily picked 7; how many remained on the branch?

7. There were 14 goblets and 12 cups on a shelf; how many more goblets were there than cups?

8. There are 7 red stripes and 6 white stripes on a United States flag; how many stripes are there in all?

9. If 2 red stripes and 1 white stripe are torn away, how many stripes remain?

10. After a baker had sold 3 rolls, he had 11 rolls left; how many had he at first?

11. William had 16 cents, and Henry had 13; how many cents less had Henry than William?

12. William spent 10 of his cents, and gave the rest to Henry; how many cents had Henry then?

13. Mr. Stone owes 7 dollars to one man, 4 to another, and 5 to another; how many dollars does he owe to all? If Mr. Stone pays 12 dollars, how many dollars will he still owe?

14. I bought some sugar for 12 cents, and an orange for 5 cents; how many cents did both cost? To pay for them, I gave a 10-cent piece, and paid the rest in cents; how many cents did it take?

10. 1. Name every two numbers that together make 11.

2. Name every two numbers that together make 12.
3. Name every two numbers that together make 13.
4. Name every two numbers that together make 14.
5. Name every two numbers that together make 15.
6. Name every two numbers that together make 16.
7. Name every two numbers that together make 17.
8. Name every two numbers that together make 18.
9. Name every two numbers that together make 19.
10. Name every two numbers that together make 20.
11. How many are 2 and 1 and 2 and 5, less 2?
12. How many are 2 and 3 and 7 and 4, less 8?
13. How many are 3 and 5 and 3 and 5, less 2, less 7, less 6?

11. 1. Count by 2's from 2 to 20, thus: 2, 4, 6, etc.

2. Count by 2's from 20 back to 2 again, thus: 20, 18, 16, etc.
3. Count by 2's from 1 to 19, and back to 1 again, thus: 1, 3, 5, etc.; 19, 17, 15, etc.
4. Count by 3's from 3 to 18, and back to 3.
5. Count by 3's from 1 to 16, and back to 1.
6. Count by 3's from 2 to 17, and back to 2.
7. Count by 4's from 4 to 20, and back to 4.
8. Count by 4's from 1 to 17, and back to 1.
9. Count by 4's from 2 to 18, and back to 2.
10. Count by 4's from 3 to 19, and back to 3.
11. Count by 5's from 5 to 20, and back to 5.
12. Count by 5's from 1 to 16, and back to 1.
13. Count by 5's from 2 to 17, and back to 2.
14. Count by 5's from 3 to 18, and back to 3.
15. Count by 5's from 4 to 19, and back to 4.

☞ For Slate Exercises in Addition, see pages 154, 155.

Section IV.

The Writing of Numbers from Twenty to One Thousand, with Addition and Subtraction.

12. The following illustrates the method of expressing numbers from twenty to one thousand:

Twenty, . . .	or 2 tens and no units	(is expressed thus),	20
Twenty-one, . .	or 2 tens and 1, ...	" .	21
Twenty-two, . .	or 2 tens and 2, ...	" .	22
Twenty-three, . .	or 2 tens and 3, ...	" .	23
Twenty-four, . .	or 2 tens and 4, ...	" .	24
Twenty-five, . .	or 2 tens and 5, ...	" .	25
Twenty-six, . .	or 2 tens and 6, ...	" .	26
Twenty-seven, .	or 2 tens and 7, ...	" .	27
Twenty-eight, .	or 2 tens and 8, ...	" .	28
Twenty-nine, . .	or 2 tens and 9, ...	" .	29
Thirty, ...	or 3 tens, ...	" .	30
Thirty-one, . . .	or 3 tens and 1, ...	" .	31
Forty, ...	or 4 tens, ...	" .	40
Forty-one, . . .	or 4 tens and 1, ...	" .	41
Fifty, ...	or 5 tens, ...	" .	50
Sixty, ...	or 6 tens, ...	" .	60
Seventy, ...	or 7 tens, ...	" .	70
Eighty, ...	or 8 tens, ...	" .	80
Ninety, ...	or 9 tens, ...	" .	90
One hundred, . .	or 10 tens, ...	" .	100
One hundred one,	or 1 hundred, no tens, and 1,	" .	101
One hundred ten,	or 1 hundred and 1 ten,	" .	110
One hundred twenty-four,	. . . .	" .	124
Two hundred,	. . . .	" .	200
Nine hundred ninety-nine,	. . . .	" .	999
One thousand,	or 10 hundreds, ...	"	1000

☞ For the writing of larger numbers, see pages 150, 153.

13. 1. How many tens are there in 20? in 30? in 40? in 60? in 80? in 100?

2. How many tens and units are there in 22? in 26? in 33? in 45? in 58? in 64? in 72? in 99?

3. How many hundreds, tens, and units, in 121? in 102? in 175? in 170? in 189? in 236? in 324?

14. 1. There are 10 sheep in one pasture and 10 in another; how many are there in both?

2. A man paid 20 dollars for his coat and 10 dollars for his pantaloons; how many dollars did he pay for both?

3. Mary spent 40 cents for an apron and 10 cents for trimmings; how many cents did she spend?

4. In an orchard there are 40 pear-trees and 20 plum-trees; how many trees are there?

5. Dora spent 50 cents for a book and 9 cents for paper; how many cents did she spend?

6. Blake spent 50 cents for fish lines, 20 cents for hooks, and 7 cents for bait; how many cents did he spend for all?

7. In a farm-yard there were 10 turkeys, 30 geese, 20 ducks, and 8 hens; how many fowls were there?

8. Margaret had 30 eggs in a basket, but she fell and broke 10 of them; how many remained whole?

9. A man bought a lot of butter for 40 dollars, but it being damaged, he sold it for 20 dollars less than he gave; what did he receive for it?

10. In a school of 50 scholars, all but 10 have perfect lessons; how many have perfect lessons?

11. A man is 56 years old, and his wife is 6 years younger than he; how old is his wife?

12. A grocer bought 40 squashes of one man and 50 of another; how many squashes did he buy? If he sold 20 squashes, how many had he left?

13. A woman, who had 80 cents, spent 10 cents for ginger and 20 cents for soda; how many cents did she spend? How many cents had she left?

14. 20 boards were sawed out of one log, 20 out of another, 40 out of another, and 30 out of another; how many boards were sawed in all?

15. 10 of the above mentioned boards were sold to one man, 50 to another, and the rest to Mr. Page; how many were sold to Mr. Page?

16. A man bought some beef for 40 cents and some mutton for 40 cents, and gave in payment one dollar (100 cents); how many cents should he receive in return?

15. What is Addition?

A cross, made thus, +, is the sign of addition. 6 + 4 means 6 and 4, and is read "6 plus 4."

What is Subtraction?

A short line, made thus, —, is the sign of subtraction. 8 — 5 means 8 less 5, and is read, "8 minus 5."

This sign, =, means equal to, or equals.

A character, made thus, $, means the same as the word dollar or dollars. $5 means 5 dollars.

1. How many are 10 and 2? 20+2? 50+2?
2. How many are 10 and 3? 30+3? 60+3?
3. How many are 10 and 6? 40+6? 80+6?
4. How many are 20 and 9? 60+9? 90+9?
5. How many are 10 and 10? 20+10? 90+20?
6. How many are 40 and 30? 60+10? 50+30?
7. How many are 20 and 60? 20+70? 40+50?
8. How many are 20 less 10? 40—10? 40—30?
9. How many are 30 less 20? 60—50? 70—60?
10. How many are 90 less 70? 100—90?
11. How many are 100 less 80? 100—50?

☞ For Slate Exercises in Addition, see pages 155, 156.

Section V.

ADDITION:—UNITS ADDED TO NUMBERS CONSISTING OF TENS AND UNITS.

16. 1. Isaac sold 9 peaches to one person, and 2 peaches to another; how many did he sell? For the first lot he received 19 cents, and for the second lot 2 cents; how many cents did he receive?

2. How many are 9 and 2? 19 and 2? 29 and 2? 39 and 2? 49 and 2? 59 and 2?

3. Otis is 9 years old, his brother Roger is 19; how old will Otis be in 3 years? How old will Roger be?

4. How many are 9 and 3? 19 and 3? 29 and 3? 39 and 3? 49 and 3? 59 and 3?

5. How many are 9 and 4? 19 and 4? 29 and 4?

6. How many are 9 and 5? 19 and 5? 39 and 5?

7. How many are 9 and 6? 19 and 6? 39 and 6?

8. How many are 9 and 7? 19 and 7? 49 and 7?

9. How many are 9 and 8? 19 and 8? 59 and 8?

10. How many are 9 and 9? 19 and 9? 79 and 9?

11. How many are 9 and 10? 19 and 10? 49 and 10?

12. Marcia bought 8 spools of cotton for 18 cents, and 3 spools for 3 cents; how many spools did she buy? How many cents did she pay for what she bought?

13. How many are 8 and 3? 18 and 3? 28 and 3?

14. How many are 8 and 4? 18 and 4? 38 and 4?

15. How many are 8 and 5? 18 and 5? 48 and 5?

16. How many are 8 and 6? 18 and 6? 58 and 6?

17. How many are 8 and 7? 28 and 7? 48 and 7?

18. How many are 8 and 8? 28 and 8? 78 and 8?

19. How many are 8 and 9? 38 and 9? 58 and 9?

20. How many are 8 and 10? 48 and 10? 68 and 10?

21. Simeon has 7 marbles, and Silas has 17; if each boy receives 4 marbles more, how many marbles will Simeon then have? How many will Silas have?

22. How many are 7 and 4? 17 and 4? 27 and 4?

23. How many are 7 and 5? 27 and 5? 47 and 5?

24. How many are 7 and 6? 37 and 6? 87 and 6?

25. How many are 7 and 7? 47 and 7? 27 and 7?

26. How many are 7 and 8? 37 and 8? 57 and 8?

27. How many are 7 and 9? 57 and 9? 37 and 9?

28. How many are 7 and 10? 47 and 10? 77 and 10?

29. A man bought a hat for $6 and a coat for $16: he afterwards sold them and gained $5 on each; how many dollars did he receive for the hat? for the coat?

30. How many are 6 and 5? 16 and 5? 36 and 5?

31. How many are 6 and 6? 16 and 6? 46 and 6?

32. How many are 6 and 7? 26 and 7? 56 and 7?

33. How many are 6 and 8? 36 and 8? 66 and 8?

34. How many are 6 and 9? 56 and 9? 36 and 9?

35. How many are 6 and 10? 76 and 10? 86 and 10?

36. Mr. Archer rode 5 miles in the cars, and Mr. Barton rode 15 miles; each afterwards walked 6 miles; how many miles did Mr. Archer go in all? How many miles did Mr. Barton go?

37. How many are 5 and 6? 15 and 6? 25 and 6?

38. How many are 5 and 7? 35 and 7? 25 and 7?

39. How many are 5 and 8? 45 and 8? 35 and 8?

40. How many are 5 and 9? 65 and 9? 85 and 9?

41. How many are 5 and 10? 75 and 10? 55 and 10?

42. William has 4 sheep, Amos has 14: if 7 should be added to the number of each, how many would each then have?

43. How many are 4 and 7? 14 and 7? 24 and 7?

44. How many are 4 and 8? 34 and 8? 54 and 8?

45. How many are 4 and 9? 44 and 9? 34 and 9?

46. How many are 4 and 10? 64 and 10? 74 and 10?

47. One lamp is worth $3, and another is worth $13; each has a shade worth $8; what arê the first lamp and shade together worth? the second?

48. How many are 3 and 8? 13 and 8? 43 and 8?

49. How many are 3 and 9? 43 and 9? 83 and 9?

50. How many are 3 and 10? 63 and 10? 53 and 10?

51. Last year there were 2 houses on Winter Street, and 12 on Pleasant Street: if 9 houses have been added to the number on each street, how many are there now on Winter Street? how many on Pleasant Street?

52. How many are 2 and 9? 52 and 9? 72 and 9?

53. How many are 2 and 10? 32 and 10? 62 and 10?

54. When Reuben was 1 year old, his mother was 31 years old; how old was each 10 years afterwards?

55. How many are 1 and 10? 11 and 10? 81 and 10?

17. 1. A man bought a saw for $4, a piece of rope for $9, and a harrow for $6; what did he pay for all?

2. A farmer sold 4 cords of pine wood for $12, and 1 cord of oak wood for $9; how many cords did he sell? How many dollars did he receive for all the wood?

3. Mr. Allen sold 3 barrels of peas for $7, and 8 barrels for $15; how many barrels did he sell? How many dollars did he receive?

4. Walter had in his library 21 books: he afterwards bought 7, and his father gave him 5; how many books had he then?

5. Julia is 15 years old, her father is 45; how old will each of them be in 5 years? in 7 years?

6. A merchant paid $19 for a chest of tea, $6 for a box of oranges, and $5 for a box of lemons; how many dollars did he pay for all?

7. At a town-meeting 35 votes were cast for one candidate, 10 for another, and 9 for another; how many votes were cast for all?

8. On board a steamer as passengers were 8 gentlemen, 9 ladies, and 9 children; how many passengers were there in all?

9. In a book there were 36 pictures of animals, 8 pictures of trees, and 7 pictures of flowers; how many pictures were there in all?

10. A man bought a cow for $65: he paid $8 for pasturing her, and then sold her for $75; how much did he gain?

11. Count by 6's from 6 to 72; from 1 to 67; from 2 to 68; from 3 to 69; from 4 to 70; from 5 to 71.

12. Count by 7's from 7 to 84; from 1 to 78; from 2 to 79; from 3 to 80; from 4 to 81; from 5 to 82; from 6 to 83.

☞ For Slate Exercises in Addition, see pages 157, 158.

Section VI.

SUBTRACTION: — UNITS TAKEN OUT OF NUMBERS CONSISTING OF TENS AND UNITS.

18. 1. Ida had 11 white roses and 21 red roses: she gave 2 of each kind to her teacher; how many white roses had she left? how many red roses?

2. How many will remain if you take 2 out of 11? 2 out of 21? 12 out of 21? 2 out of 31? 2 out of 41?

3. How many cents will remain, if you take 3 cents out of 11 cents? 3 cents out of 21 cents? 13 cents out of 21 cents? 3 cents out of 31 cents? 4 cents out of 11 cents? 4 cents out of 21 cents? 4 cents out of 41 cents?

4. How many dollars will remain, if you take 5 dollars out of 11 dollars? 5 dollars out of 31 dollars? 5 dollars out of 41 dollars? 6 dollars out of 11 dollars? 6 dollars out of 21 dollars? 6 dollars out of 41 dollars? 6 dollars out of 51 dollars?

5. How many will remain, if you take 7 out of 11? 7 out of 21? 17 out of 21? 8 out of 11? 8 out of 21? 8 out of 31? 8 out of 51?

6. How many will remain, if you take 9 out of 11? 9 out of 21? 19 out of 21? 9 out of 61? 10 out of 11? 10 out of 21? 10 out of 81?

7. A market-woman had 12 melons, and 22 cabbages: if she sold 3 of each to one person, how many melons had she left? how many cabbages?

8. How many are 12 less 3? 22 less 3? 22 less 13? 42 less 3? 12 less 4? 32 less 4? 62 less 4?

9. How many are 12 less 5? 22 less 5? 22 less 15? 12 less 6? 32 less 6? 42 less 6? 92 less 6?

10. How many are 12 less 7? 22 less 7? 22 less 17? 52 less 7? 12 less 8? 22 less 8? 72 less 8?

11. How many are 12 less 9? 22 less 9? 22 less 19? 62 less 9? 22 less 10? 82 less 10? 92 less 10?

12. A man owed $13 to one man, and $23 to another: he paid $4 to each man; how much did he still owe to the first man? how much to the second?

13. How many are 13 less 4? 23 less 4? 23 less 14? 33 less 5? 43 less 5?

14. How many are 23 less 6? 53 less 6? 53 less 7? 63 less 7?

15. How many are 23 less 8? 43 less 8? 23 less 9? 53 less 10?

16. Thomas is 14 years old, Willard is 24 years old, and Mary is 5 years old; how many years older is Thomas than Mary? is Willard than Mary?

17. How many more are 14 than 5? 24 than 5? 24 than 15? 34 than 5? 34 than 6? 34 than 7?

18. How many more are 14 than 8? 24 than 18? 14 than 9? 24 than 9? 24 than 19? 44 than 9?

19. A father is 45 years old, and his son is 6 years old; how old was the father when his son was born?

20. How many are 45 less 6? 45 less 7? 55 less 8?

21. How many are 16 less 7? 26 less 7? 36 less 8?

22. How many are 17 less 8? 27 less 8? 27 less 18?

23. How many are 18 less 9? 38 less 9? 68 less 9?

24. How many are 102 less 9? 107 less 10?

19. 1. A baker had 24 loaves of bread: he sold 10 loaves to one man; how many had he left? He sold 4 loaves to another, and kept the rest himself; how many loaves did he keep?

2. An orphan asylum had 28 children: if 9 were sent to the West, how many remained? If 11 of the number that remained, afterwards went to learn trades, how many were then at the asylum?

3. Mr. Williams paid $25 for three buffalo robes: for one he paid $7, for another $10; how much did he pay for the third?

4. A merchant had 54 pieces of sheeting: he sold 10 pieces to one person, 7 pieces to another, and 6 pieces to another; how many pieces had he left after each sale?

5. A man had $60, and spent $20 for a coat, $30 for an overcoat, and $5 for a pair of boots; how many dollars had he left?

6. Subtract by 6's from 72, thus: 72, 66, 60, 54, etc.

7. Subtract by 6's from 71 to 5; from 70 to 4; from 69 to 3; from 68 to 2; from 67 to 1.

8. Subtract by 7's from 84 to 7; from 83 to 6; from 82 to 5; from 81 to 4; from 80 to 3; from 79 to 2; from 78 to 1.

☞ For Slate Exercises in Subtraction, see pages 159, 160.

Section VII.

ADDITION AND SUBTRACTION OF NUMBERS, CONSISTING OF TENS AND UNITS.

20. 1. There were 32 pigeons on one tree and 7 on another; how many pigeons were on both trees? 8 of the pigeons flew away; how many were left?

2. There were 14 planks sawed out of one log, 8 out of another, and 9 out of a third; how many planks were there in all? 7 of these planks were used in repairing a bridge; how many were left?

3. In a farm-yard were 8 hens, 21 chickens, and 5 turkeys; how many fowls were in the yard? A hawk took away 3 of the chickens, and a fox carried off 4 hens; how many fowls were left?

4. On the third day of July, John bought some fire-crackers for 12 cents, some pin-wheels for 10 cents, a bunch of serpents for 9 cents, and a slow-match for 6 cents; how much did he pay for all?

5. A drover, who had 29 sheep, has sold 6 of them; how many has he left? How many must he now buy to have 30?

6. There are 26 fowls in a yard: 3 are ducks, 8 are geese, and the rest hens; how many are hens?

7. A man picked in one pasture 19 quarts of berries, in another 7, and in another 5: he sold 4 quarts to one person, and 3 to another; how many quarts had he left?

8. Frank has bought a slate for 30 cents, a pen for 6 cents, some ink for 6 cents, and an inkstand for 6 cents: if he gives a 50-cent piece in payment, how many cents should be returned to him?

9. Mrs. Gay bought a gold watch for $78, a card-case for $5, a breast-pin for $8, and a ring for $4: if she gave two 50-dollar bills to pay for them, how many dollars should she receive in return?

10. On a certain day, Mrs. Jones earned 46 cents by making two shirts, 5 cents by hemming a handkerchief, and 9 cents by making some button-holes; how much did she earn in all?

11. She spent 15 cents for bread, 2 cents for milk, 6 cents for herrings, and 7 cents for potatoes; how much money had she left for other expenses?

12. It took Mrs. Jones 13 hours to make the shirts, and 2 hours to do the rest of the sewing: there are 24 hours in the day; how many hours had she for rest and other duties?

21. 1. How many are 18+6+5+4?

2. How many are 2+47+8+3?

3. How many are 35+2+6+4+8?

4. How many are 39+5+8+7+4+6?

5. How many are 43+5+4+8+7+9?

6. Count by 8's from 8 to 96 and back again; from 1 to 89; from 2 to 90; from 3 to 91; from 4 to 92; from 5 to 93; from 6 to 94; from 7 to 95.

7. Count by 9's from 9 to 108 and back; from 1 to 100; from 2 to 101; from 3 to 102; from 4 to 103; from 5 to 104; from 6 to 105; etc.

22. 1. Ellen spent 24 cents for a book, and 16 cents for paper; how many cents did she spend?

2. How many are 24 and 16?

Solution. 24 and 10 are 34, and 6 more are 40; therefore, 24 and 16 are 40.

3. How many are 36 and 14? 25+15?

4. How many are 22 and 18? 57+12?

5. How many are 23 and 19? 54+17?

6. How many are 79 and 16? 68+23?

7. How many are 27 and 15? 17 and 25? 27 and 25?

8. How many are 39 and 12? 19 and 32? 39 and 42?

9. Mr. Fay rode 35 miles on the railroad, 26 miles in a stage, and 15 miles in his own carriage; how many miles did he ride in all?

10. How many are 35 and 26 and 15?

Solution. 35 and 20 are 55, and 6 more are 61; 61 and 10 are 71, and 5 more are 76: therefore, etc.

11. How many are 23+26+14?

12. How many are 37+26+22?

13. How many are 19+33+42?

14. How many are 18+47+48?

15. A tailor paid $17 for some cloth and trimmings, and made them into a coat which he sold for $35; how many dollars did he get for his work?

16. How many are 35 less 17?

Solution. 35 less 10 are 25, less 7 are 18.

17. How many are 29 less 15? 48—15?

18. How many are 48 less 26? 58—19?

19. How many are 44 less 18? 44—28?

20. How many are 71 less 34? 81—24?

21. How many are 43 less 17? 44—29?

22. How many are 64 less 28? 68—39?

23. 1. Sarah, Mary, and Ellen trimmed their school-room with evergreen: Sarah wove 12 yards for festoons, Mary wove 30 yards, and Ellen wove 7 yards; how many yards did all weave?

2. Sarah used 20 yards of twine, Mary used 17 yards, and Ellen used 6 yards of one kind and 9 of another; how many yards of twine did all use?

3. John and Emery made a collection of beetles: John collected 47 different kinds, and Emery 29 different kinds; how many more kinds did John collect than Emery?

4. Charles and Mary made a collection of the different kinds of grasses: Mary found 21 different kinds, and Charles found 18 different kinds, 15 of which were like some of Mary's; how many different kinds did both find?

5. Alvin and Reuben hired a boat for 75 cents: Reuben paid 37 cents of the money, and Alvin paid the rest; how much did Alvin pay?

6. They gathered 87 pond-lilies, 50 of which they sold at 1 cent apiece; how many cents did they receive for them? How many lilies remained unsold?

7. They gave 12 of the lilies that remained to Annie, and the rest to Reuben's sick brother; how many did they give to Reuben's sick brother?

8. They caught 9 perch, 12 carp, 7 pickerel, and 8 eels; how many fishes did they catch? They threw away 13 of the fishes, gave 12 to Alvin's mother, and what remained to widow Dale; how many did they give widow Dale?

9. Henry gathered 102 pond-lilies, 9 of which he gave to Mary, and 8 to Charles; how many had he left? Of these 37 were in full bloom, and the rest were buds; how many were buds?

10. How much less than 90 is 47 plus 18?

11. Man has 32 teeth in a full set: of these 8 are incisors, 4 are cuspids, 8 are bi-cuspids, and the rest are grinders; how many are grinders?

12. 49 scholars belong to Mr. Boyd's school: on a certain day 17 were dismissed for good lessons, and 4 were absent; how many were then in the school-room?

13. If in a gale 28 feet of the upper part of a tree was blown off, and the part which remained was 74 feet in height, what was the height of the tree before the gale?

14. There are 92 days in the summer months: June has 30 days, July 31 days, and August the remainder; how many days has August?

15. Frank sold 9 apples for 18 cents, 20 for 15 cents, and 8 for 8 cents; how many apples did he sell? How much money did he receive for them?

16. With this money he bought a hat for 25 cents, and a basket for 12; how much money had he left?

17. Horace wanted a drawing-book worth 37 cents, a pencil worth 6 cents, and a ruler worth 18 cents: his father gave him two 25-cent pieces, two 10-cent pieces, one 5-cent piece, and one 3-cent piece, telling him to spend what money was left, after making his purchases, for paper; how much money had he to pay for paper?

18. Mr. Dana bought a dinner for 72 cents, and afterwards called for a cup of tea worth 10 cents; how much should he pay for both? He gave in payment one 50-cent piece, three 10-cent pieces, and one 5-cent piece; how much change should be returned to him?

☞ For Slate Exercises, see pages **161, 162**.

Section VIII.

MULTIPLICATION TABLES.

24. The Multiplication Tables are inserted here for future reference; they may be learned in connection with the lessons which follow.

1	**1** is 1	**2** is 2	**3** is 3	**4** is 4
2	1's are 2	2's are 4	3's are 6	4's are 8
3	1's " 3	2's " 6	3's " 9	4's " 12
4	1's " 4	2's " 8	3's " 12	4's " 16
5	1's " 5	2's " 10	3's " 15	4's " 20
6	1's " 6	2's " 12	3's " 18	4's " 24
7	1's " 7	2's " 14	3's " 21	4's " 28
8	1's " 8	2's " 16	3's " 24	4's " 32
9	1's " 9	2's " 18	3's " 27	4's " 36
10	1's " 10	2's " 20	3's " 30	4's " 40
11	1's " 11	2's " 22	3's " 33	4's " 44
12	1's " 12	2's " 24	3's " 36	4's " 48

1	**5** is 5	**6** is 6	**7** is 7	**8** is 8
2	5's are 10	6's are 12	7's are 14	8's are 16
3	5's " 15	6's " 18	7's " 21	8's " 24
4	5's " 20	6's " 24	7's " 28	8's " 32
5	5's " 25	6's " 30	7's " 35	8's " 40
6	5's " 30	6's " 36	7's " 42	8's " 48
7	5's " 35	6's " 42	7's " 49	8's " 56
8	5's " 40	6's " 48	7's " 56	8's " 64
9	5's " 45	6's " 54	7's " 63	8's " 72
10	5's " 50	6's " 60	7's " 70	8's " 80
11	5's " 55	6's " 66	7's " 77	8's " 88
12	5's " 60	6's " 72	7's " 84	8's " 96

1	**9** is 9	**10** is 10	**11** is 11	**12** is 12
2	9's are 18	10's are 20	11's are 22	12's are 24
3	9's " 27	10's " 30	11's " 33	12's " 36
4	9's " 36	10's " 40	11's " 44	12's " 48
5	9's " 45	10's " 50	11's " 55	12's " 60
6	9's " 54	10's " 60	11's " 66	12's " 72
7	9's " 63	10's " 70	11's " 77	12's " 84
8	9's " 72	10's " 80	11's " 88.	12's " 96
9	9's " 81	10's " 90	11's " 99	12's " 108
10	9's " 90	10's " 100	11's " 110	12's " 120
11	9's " 99	10's " 110	11's " 121	12's " 132
12	9's " 108	10's " 120	11's " 132	12's " 144

Section IX.

MULTIPLICATION OF THE NUMBERS 2, 3, 4, AND 5.

25. 1. One leaf of a book has 2 pages; how many pages have 3 leaves?

2. How many are 2 and 2 and 2, or 3 times 2?

3. At 2 dollars each, what will 4 chairs cost?

4. How many are 2+2+2+2, or 4 times 2?

5. At 2 cents each, what will 5 lemons cost? what will 6 lemons cost? 7 lemons?

6. Add by 2's from 2 to 24.

7. Repeat Table **2**, page 33.

8. How many are 2 times 2? 7 times 2? 9 times 2? 12 times 2? 8 times 2? 11 times 2? 10 times 2?

9. If a traveller can go 3 miles in an hour, how many miles can he go in 2 hours? in 3 hours?

10. How many are 3+3+3, or 3 times 3?

11. At 3 dollars a pair, how many dollars must be paid for 4 pairs of boots? for 5 pairs?

12. Add by 3's from 3 to 36.

13. Repeat Table **3**, page 33.

14. How many are 8 times 3? 6 times 3? 10 times 3? 9 times 3? 11 times 3? 7 times 3? 12 times 3?

15. How many pounds of butter are there in 2 boxes, if there are 4 pounds in each box?

16. 2 times 4 are how many?

17. How many desks are there in 3 rows, if there are 4 desks in each row?

18. Add by 4's from 4 to 48.

19. Repeat Table **4**, page 33.

20. How many are 6 times 4? 8 times 4? 4 times 4? 12 times 4? 9 times 4? 5 times 4? 11 times 4?

21. What cost 2 oranges at 4 cents apiece? What cost 4 oranges at 2 cents apiece?

22. How many are two 4's? four 2's?

23. How many trees are there in 4 rows of 3 trees each? How many trees in 3 rows of 4 trees each?

24. Is there any difference between four 3's and three 4's?

25. Warren has 5 tops, and Willard has 2 times as many (or two 5's of tops); how many has Willard?

26. A farmer planted 5 kernels of corn in each hill; how many kernels did he plant in 3 hills? in 4 hills?

27. Add by 5's from 5 to 60.

28. Repeat the Table of 5's, page 33.

29. How many are five 5's? eight 5's? ten 5's? six 5's? seven 5's? nine 5's? eleven 5's?

30. How many are two 5's? five 2's? four 5's? five 4's?

26. 1. What will 7 apples cost, at 2 cents apiece?

Solution. If 1 apple costs 2 cents, 7 apples will cost seven 2's of cents: seven 2's of cents are 14 cents; therefore, 7 apples, at 2 cents apiece, will cost 14 cents.

2. When oil is $3 a gallon, what will 7 gallons cost?

3. There are 4 quarts in 1 gallon; how many quarts are there in 7 gallons? in 9 gallons? in 12 gallons?

4. There are 3 feet in 1 yard; how many feet are there in 6 yards? in 8 yards? in 9 yards? in 11 yards?

5. What will 5 quarts of milk cost at 5 cents a quart?

6. If a horse can travel 1 mile in 5 minutes, in how many minutes can he travel 12 miles?

7. What cost 11 cords of wood at $2 a cord? at $4 a cord? at $1? at $5?

8. If a man earns $2 a day, and his team earns $2 a day, how much do both earn in 1 day? in 7 days?

Section X.

MULTIPLICATION OF THE NUMBERS 6, 7, 8, AND 9.

27. 1. There are 6 working days in 1 week; how many working days are there in 2 weeks? in 3 weeks? in 4 weeks?

2. Add by 6's from 6 to 72.

3. Repeat the Table of **6**'s, page 33.

4. How many are eight 6's? eleven 6's? nine 6's? seven 6's? ten 6's? five 6's? six 5's?

5. If a horse eats 7 quarts of oats in 1 day, how many quarts will he eat in 2 days? in 3 days? in 4 days?

6. Add by 7's from 7 to 84.

7. Repeat the Table of **7**'s, page 33.

8. How many are eleven 7's? seven 7's? nine 7's? twelve 7's? eight 7's? four 7's? seven 4's? five 7's? seven 5's?

9. There are 8 quarts in 1 peck; how many quarts are there in 2 pecks? in 3 pecks? in 4 pecks?

10. Add by 8's from 8 to 96.

11. Repeat the Table of **8**'s, page 33.

12. How many are five 8's? eight 8's? six 8's? eight 6's? twelve 8's? nine 8's? seven 8's? eight 7's?

13. At $9 each, what will be the cost of 2 saddles? of 3 saddles? of 4 saddles?

14. Add by 9's from 9 to 108.

15. Repeat the Table of **9**'s, page 33.

16. How many are five 9's? eight 9's? six 9's? twelve 9's? nine 9's? seven 9's? nine 7's? four 9's? nine 4's?

28. 1. Jacob can buy 6 marbles for 1 cent; how many marbles can he buy for 5 cents? Why?

2. At 7 cents a bunch, what cost 6 bunches of beets? 8 bunches?

3. In a class there were 8 boys, and 2 times as many girls (or two 8's of girls); how many girls were there?

4. In a hop-field there were twelve rows of hops, and 9 hills in each row; how many hills were there?

5. If a horse-car goes 9 miles in an hour, and a rail-car goes four times as far (or four 9's of miles), how many miles does the rail-car go in an hour?

6. If the interest of $1 for a year is 6 cents, what will be the interest of the same sum for 2 years? for 3 years? for 4 years?

7. If the interest of $1 for a year is 7 cents, what is the interest of $5 for the same time? of $12?

8. If 2 boys can weed a garden in 6 days, in how many days can 1 boy weed it?

9. If 3 men can do a piece of work in 6 days, in how many days can 1 man do it?

10. If a family of 6 boys eat a bushel of apples in 7 days, how many days would a bushel last 1 boy at the same rate?

11. If a man buys pineapples at 8 cents apiece, and sells them at 15 cents apiece, what does he gain on 1 pineapple? on 4 pineapples?

12. If a woman earns $10 in a week, and spends all but $3, how many dollars does she spend in 1 week? in 7 weeks?

13. If it requires 1 yard to make a vest, and 7 yards to make a cloak, how many yards will be required to make 5 vests and 5 cloaks?

14. John and James hire a boat by the hour: if John pays 4 cents an hour, and James pays 5 cents, how much do both pay for 2 hours? for 7 hours?

Section XI.

MULTIPLICATION OF THE NUMBERS 10, 11, AND 12.

92. 1. At 10 cents each, what must be paid for 2 candles? for 3 candles? for 4 candles?

2. Add by 10's from 10 to 120.

3. Repeat the Table of **10**'s, page 33.

4. How many are ten 10's? six 10's? nine 10's? twelve 10's? eight 10's? ten 8's?

5. At 11 cents a dozen, what will 2 dozen eggs cost? 3 dozen? 4 dozen? 5 dozen?

6. Add by 11's from 11 to 132.

7. Repeat the Table of **11**'s, page 33.

8. How many are six 11's? eleven 11's? twelve 11's? eight 11's? ten 11's? eleven 10's?

9. If hay is worth $12 a ton, how many dollars must be paid for 2 tons? for 3 tons? for 4 tons?

10. Add by 12's from 12 to 144.

11. Repeat the Table of **12**'s, page 33.

12. How many are eight 12's? ten 12's? eleven 12's? five 12's? nine 12's? seven 12's? twelve 7's?

13. How many yards of calico in 6 dresses of 10 yards each? of 11 yards each?

14. If 11 plums will pay for 1 apple, how many plums will pay for 7 apples? for 11 apples?

15. If 12 yards of carpeting which is 2 yards wide, will cover a floor, how many yards that is 1 yard wide will cover the same floor?

16. If a quantity of bread will last 12 persons 6 days, how many persons would the same bread last 1 day?

☞ For Slate Exercises in Multiplication, see pages 163 to 165.

Section XII.

MISCELLANEOUS REVIEW.

30. In performing each of the examples in the last four sections, we have united a number of equal numbers; such a process is **Multiplication.**

A cross, made thus, $\times$, is the sign of Multiplication. $5\times4=20$ means four 5's equal 20; and is read "four 5's equal 20," "5 multiplied by 4 equals 20," or "4 times 5 equals 20."

NOTE. In teaching the subject of multiplication, either of the expressions "four 5's equal 20," or "5 multiplied by 4 equals 20," is preferable to "four times 5 equals 20," since, in the operation, we do not take 1 five four times, but 4 fives at once. When the pupil has become familiar with the subject, however, he may use either of these three forms of expression.

1. How many are 6 multiplied by 7? 6×12?
2. How many are eight 7's? 7×9? 7×12?
3. How many are six 8's? 8×7? 8×9?
4. How many are four 9's? 9×7? 9×6?
5. How many are two 4's, plus 2? two 6's, plus 4?
6. How many are five 4's, plus 9? five 9's, plus 9?
7. How many are seven 5's, plus 8? seven 8's, plus 7?
8. How many are twelve 8's, plus 11?
9. How many are eleven 7's, plus 12?
10. How many are twelve 8's, plus 12?
11. How many are ten 11's, plus 15?
12. How many are twelve 7's, plus 20?
13. How many are 4×8, plus 6×2, less 3×3?
14. How many are 5×9, plus 3×5, less 4×4?
15. How many are 8×7, plus 7×2, less 4×7?
16. How many are 9×7, plus 4×3, less 6×8?
17. How many are 7×11, less 8×9, plus 2×7?
18. How many are 10×9, less 4×10, plus 8×7?
19. How many are 8×11, less 5×10, plus 6×9?
20. How many are 8×9, less 4×4, plus 12×6?

31. 1. What will 10 yards of cloth cost at $11 a yard? at $12?

2. How much should you pay for 2 rabbits at 12 cents apiece, and 2 doves at 10 cents apiece?

3. Philo had 50 cents: with this money he bought 4 pounds of figs at 12 cents a pound; how many cents had he left?

4. I bought 3 barrels of flour at $11 a barrel, and sold the whole for $40; how many dollars did I gain?

5. If by working 10 hours a day a man can build a fence in 3 days, in how many days can he build it by working 1 hour a day?

6. John can saw a cord of wood in 5 days by working 2 hours each day, but he wishes to saw it in 1 day; how many hours must he work on that day?

7. Two men start from the same place and travel in opposite directions: one travels 3 miles an hour, and the other 4 miles an hour; how far apart will they be at the end of 1 hour? of 2 hours? of 5 hours?

8. Two men start from the same place, and travel in the same direction, one at the rate of 2 miles an hour, and the other at the rate of 12 miles an hour; how far apart will they be in 1 hour? in 2 hours? in 4 hours?

9. A person bought 8 pencils at 6 cents apiece, and 4 quires of paper at 9 cents a quire; what did he pay for the pencils? for the paper? for both?

10. A person bought 6 quarts of vinegar at 7 cents a quart, and 3 pounds of sugar at 9 cents a pound; what did he pay for both?

11. Abby bought 4 pounds of codfish at 9 cents a pound, and 2 mackerel at 8 cents apiece; how much did she pay for what she bought?

12. What cost 4 heads of lettuce at 10 cents a head, 6 pounds of tomatoes at 5 cents a pound, and 2 pounds of beef at 11 cents a pound?

13. A merchant had 7 barrels of flour: he sold 5 barrels at $12 a barrel, and the remainder at $11 a barrel; how much did he receive for the lot?

14. Laura earns 72 cents in a day, Ada earns 60 cents in a day; how much more does Laura earn in 1 day than Ada? how much more in 9 days?

15. Josiah sold 2 quarts of raspberries to Mrs. Lamb for 27 cents, and 3 quarts to Mrs. Hewins at 11 cents a quart; how many cents did he receive for his raspberries?

16. With the money which he received, he paid 20 cents for a basket, and bought 3 apples at 2 cents apiece; how many cents had he left?

17. William is 11 years old: his father is 4 times as old, and 9 years more; how old is his father?

18. Arthur bought 7 pounds of dates at 12 cents a pound, and had 7 cents left; how many cents had he at first?

19. Mary has 37 flowers: if she makes 3 bouquets, putting 12 flowers into each, how many flowers will she have left?

20. Joseph sold 5 quarts of berries at 10 cents a quart, and received in payment 4 pounds of rice at 9 cents a pound, and the rest in money; how much money did he receive?

21. A merchant bought 6 barrels of pork at $12 a barrel, and gave in payment 5 cords of wood at $11 a cord, and the rest in money; how many dollars did he pay in money?

32. 1. Four 6's, less 5, less 8, are how many?

2. Two 3's, multiplied by 8, less 7, are how many?

3. Nine 5's, plus 3, plus 5, less 40, are how many?

4. 6 multiplied by 6, plus 4, plus 25, are how many?

5. Multiply 3 by 5, add 2, add 3, subtract 2, subtract 10, multiply by 5, add 15; how many have you?

6. Take 1, add 5, add 7, subtract 8, multiply by 11, subtract 5, subtract 5, subtract 20, subtract 20, add 1, multiply by 9; how many have you?

7. Think of any number less than 7, multiply that number by 5, subtract 3 numbers equal to the number thought of, add 2, add 4, take away 5, subtract two numbers equal to the number thought of; what number have you?

NOTE. The teacher can excite a lively interest in combining numbers rapidly, by giving the pupils impromptu exercises similar to the above.

33. 1. How many are four 23's?

Solution. 23 equals 20 plus 3; four 20's are 80, and four 3's are 12; which, added to 80, make 92.

2. How many are two 13's? 22's? 35's? 46's?
3. How many are three 21's? 18's? 36's? 40's?
4. How many are four 22's? 13's? 19's? 25's?
5. How many are five 16's? 17's? 28's? 22's?
6. How many are six 13's? 17's? 19's? 21's?
7. How many are seven 14's? 20's? 17's? 19's?
8. How many are eight 13's? 16's? 18's? 19's?
9. How many are nine 15's? 14's? 19's? 18's?
10. How many are ten 14's? 16's? 13's? 17's?

NOTE. If the teacher desires to have multiples larger than those given in the Multiplication Tables, page 33, committed to memory, the pupil can construct tables for himself in the following manner, and commit them from his slate or paper: —

$2\times13=26$	$3\times13=39$	$4\times13=52$
$2\times14=28$	$3\times14=42$	$4\times14=56$
$2\times15=30$	$3\times15=45$	$4\times15=60$
etc.	etc.	etc.

☞ For Slate Exercises in Multiplication, see pages 165 to 167.

Section XIII.

DIVISION, — SUBTRACTIVE FORM.

34. 1. Charles had 6 peaches, which are just enough to give 1 peach to each of his sisters; how many sisters has he?

2. Albert can carry 4 sticks of wood into the house at a time; how many times must he go to carry in 8 sticks?

3. How many armfuls, of 4 sticks each, must be taken away to remove a pile of 16 sticks?

4. How many desks will be required to seat 10 scholars, if 2 scholars can sit at each desk?

5. If in a class there are 15 scholars sitting in rows, and in each row there are 3 scholars, how many rows are there?

6. In a procession, there were 16 boys walking in ranks of 2 boys each; how many ranks were there?

7. If the 16 boys had walked in ranks of 4 boys each, how many ranks would there have been?

8. If there are 2 oxen in a yoke of oxen, how many yoke will 18 oxen make?

9. A man can build 5 rods of fence in a day; how many days will it take him to build 20 rods?

10. How many 2's are there in 12? in 10? in 18? in 14? in 2? in 16? in 20? in 24?

11. How many 3's are there in 12? in 21? in 15? in 18? in 27? in 30? in 36? in 24?

12. How many 4's are there in 44? in 12? in 16? in 20? in 28? in 36? in 44? in 32?

13. How many 5's are there in 10? in 15? in 20? in 30? in 25? in 45? in 50? in 55?

14. How many 6's are there in 12? in 18? in 24? in 42? in 54? in 36? in 48? in 66? in 72?

15. How many 7's are there in 14? in 28? in 42? in 56? in 77? in 63? in 49? in 35? in 84?

16. How many 8's are there in 24? in 40? in 64? in 96? in 88? in 32? in 48? in 56? in 16?

17. How many 9's are there in 27? in 45? in 54? in 36? in 81? in 63? in 72? in 99? in 108?

18. How many 10's are there in 30? in 50? in 80? in 110? in 40? in 20? in 70? in 100? in 90?

19. How many 11's are there in 66? in 88? in 33? in 44? in 22? in 99? in 110? in 77? in 121?

20. How many 12's are there in 36? in 24? in 48? in 96? in 84? in 120? in 108? in 60? in 132?

35. 1. How many oranges, at 5 cents apiece, can be bought for 15 cents?

*Solution.** If 1 orange costs 5 cents, as many oranges can be bought for 15 cents as there are 5's in 15: there are three 5's in 15; therefore, 3 oranges at 5 cents apiece can be bought for 15 cents.

2. How many hats, at $7 apiece, can be bought for $21? for $35? Why?

3. At 9 cents a loaf, how many loaves of bread can be bought for 45 cents? for 108 cents?

4. How many barrels of flour, at $12 a barrel, can be bought for $60? for $96?

5. Mr. Flint gave 30 apples to some boys, giving 5 apples to each; how many boys were there?

6. To how many boys could Mr. Flint have given the apples, if he had given them 6 apples apiece?

7. If Timothy can earn $5 in a week, in how many weeks can he earn $60?

* If preferred the following solution can be given.

SOLUTION. If 5 cents will buy 1 orange, 15 cents, which is three 5's of cents, will buy 3 oranges; therefore, etc.

8. How many pounds of sugar, at 9 cents a pound, can be bought for 72 cents? How many at 8 cents?

9. If the board of a family for 1 week is $11, how many weeks' board can be obtained for $44?

10. In 1 week there are 7 days; how many weeks are there in 63 days?

11. If a steamboat can go 12 miles in one hour, in how many hours can she go 120 miles?

12. If it takes 8 yards of cloth to make a suit of clothes, how many suits can be made from 64 yards?

13. A farmer put 24 bushels of wheat into bins, each bin containing 8 bushels; how many bins were used?

14. The farmer afterwards put these 24 bushels of wheat into bags for market, putting 2 bushels in each bag; how many bags did he fill?

Section XIV.

DIVISION, — SUBTRACTIVE FORM: EXAMPLES IN MULTIPLICATION AND DIVISION COMBINED.

36. 1. 4 are how many 2's? 4's? 1's?

2. 8 are how many 2's? 4's? 8's?
3. 10 are how many 2's? 5's?
4. 14 are how many 2's? 7's?
5. 15 are how many 3's? 5's?
6. 22 are how many 2's? 11's?
7. 33 are how many 3's? 11's?
8. 35 are how many 5's? 7's?
9. 42 are how many 6's? 7's? 2's?
10. 44 are how many 4's? 11's? 22's?
11. 49 are how many 7's? 1's?
12. 54 are how many 9's? 6's? 2's?

13. 55 are how many 5's? 11's?
14. 56 are how many 7's? 8's? 28's?
15. 64 are how many 8's? 4's? 2's?
16. 66 are how many 6's? 11's? 33's?
17. 70 are how many 7's? 10's?
18. 81 are how many 9's? 27's?
19. 84 are how many 7's? 12's? 2's? 21's?
20. 88 are how many 8's? 11's? 4's? 44's?
21. 90 are how many 9's? 10's? 3's? 15's?
22. 96 are how many 8's? 12's? 24's?
23. 100 are how many 10's? 2's? 4's? 20's?
24. 108 are how many 9's? 12's?
25. 110 are how many 10's? 11's?
26. 121 are how many 11's?
27. 132 are how many 11's? 12's?
28. 144 are how many 12's? 24's?

37. 1. Albert sold 4 oranges at 3 cents apiece; how many cents did he receive for them? With the money which he received, he bought apples at 2 cents apiece; how many apples did he buy?

Solution. If 1 orange was sold for 3 cents, 4 oranges were sold for four 3's of cents, or 12 cents. He bought as many apples at 2 cents apiece as there are 2's in 12; there are six 2's in 12; therefore, he bought 6 apples.

2. Two 3's are how many 2's?

3. How many postage stamps at 2 cents each, should be given for 8 stamps at 1 cent each?

4. Eight 1's are how many 2's?

5. Myra bought 9 stamps at 2 cents each, and paid for them with 3-cent pieces; how many 3-cent pieces did she pay for them?

6. Nine 2's are how many 3's?

7. If a horse can travel 8 miles in 1 hour, how far can he travel in 2 hours? If a man can walk 4 miles

in 1 hour, in how many hours can the man walk as far as the horse can travel in 2 hours?

8. Two 8's are how many 4's?

9. What cost 5 pairs of shoes at $4 a pair? How many bills of $10 each will pay for these shoes?

10. If a man can build a fence in 5 days by working 6 hours a day, how many hours does it take him to build it? In how many days could he build it, if he should work 10 hours a day?

11. There are 12 eggs in a dozen: if it requires 3 eggs to make a pound of cake, how many pounds of cake can be made with 2 dozen eggs?

12. How many rows of trees, having 6 trees in each row, can be set from 4 rows having 12 trees in each row?

13. How many pounds of cheese at 9 cents a pound, will pay for 6 baskets at 12 cents each?

14. If a horse travels 6 miles an hour, in how many hours can he travel as far as a car can go in 3 hours, the car going at the rate of 20 miles an hour?

15. A newsboy sold 4 papers at 3 cents apiece, and 2 papers at 4 cents apiece: with the money which he received, he bought some papers at 2 cents apiece; how many papers did he buy?

16. A man spends 15 cents a day for newspapers at 5 cents apiece; how many papers does he buy in 1 day? in 6 days? in 11 days?

17. A woman has a jug which it costs 16 cents to get filled with syrup at 8 cents a quart; how many quarts does the jug hold? How many quarts will it take to fill the jug 11 times?

18. Five 12's are how many 10's?

19. Four 10's are how many 5's?

20. Three 10's are how many 6's?

21. Nine 8's are how many 12's?

22. Eight 6's are how many 4's?

Section XV.

DIVISION, — SUBTRACTIVE FORM, WITH REMAINDERS.

38. 1. In 7 shoes how many pairs of shoes, and how many shoes remain?

Solution. If there are 2 shoes in 1 pair, in 7 shoes there are as many pairs as there are 2's in 7 : there are three 2's in 7, and 1 remains; therefore, in 7 shoes there are 3 pairs, and 1 shoe remains.

2. 4 shoes will shoe 1 horse; how many horses can be shod with 10 shoes, and how many shoes will remain?

3. 10 are how many 4's, and what remains?

4. In 11 pages of history how many lessons are there of 3 pages each, and how many pages remain?

5. 11 are how many 3's? *

6. If 3 yards of cloth will make one sack, how many sacks can be made from 13 yards, and how many yards will remain?

7. 13 are how many 3's?

8. If 5 kernels of corn are put in each hill, how many hills of corn can be planted with 12 kernels, and how many kernels will remain?

9. 12 are how many 5's?

10. Robert has 15 cents; how many pencils can he buy at 4 cents apiece, and how many cents will remain?

11. In 25 eggs how many dozen are there, and how many eggs remain?

12. A man, who owed $32, paid as much of the debt as possible in 5-dollar bills, and the rest in 1-dollar bills; how many bills of each sort did it require?

13. If a basket costs 8 cents, how many baskets can be bought for 27 cents, and how many cents will remain?

* The answers to this and the remaining examples of its class in this section, should be given with remainders.

14. If a man has 34 apples, and sells them at the rate of 3 apples for 1 cent, how many cents' worth of apples can he sell, and how many apples will remain?

15. How many weeks are there in 16 days, and how many days remain?

16. How many weeks in 20 days? in 21 days?

17. 18 are how many 4's? 5's? 6's? 7's? 8's?

18. 25 are how many 2's? 3's? 4's? 5's? 6's?

19. Name the numbers, from 6 to 40, which leave no remainder when they are divided by 5.

20. What may these numbers be called?

Ans. **Multiples** of 5.

21. What are the multiples of 6, from 6 to 72?

Ans. 6, 12, 18, etc.

22. Name the multiples of 7, from 7 to 84; of 8, from 8 to 96; of 9, from 9 to 108.

23. Name the numbers, from 1 to 20, which leave a remainder when they are divided by 2.

Ans. 1, 3, 5, 7, etc.

24. Name the numbers, from 1 to 30, which leave a remainder when they are divided by 3.

25. 17 are how many 2's? 1's? 7's? 4's? 5's? 3's?

26. 21 are how many 8's? 10's? 2's? 7's? 4's? 9's? 6's? 3's? 5's?

27. 28 are how many 12's? 3's? 4's? 5's? 6's? 7's? 8's? 9's? 10's? 11's?

28. 32 are how many 9's? 7's? 3's? 5's? 8's? 6's? 4's? 10's? 11's? 12's?

29. 45 are how many 5's? 4's? 10's? 9's? 6's? 7's?

30. 50 are how many 12's? 7's? 5's? 8's? 9's? 6's?

31. 63 are how many 6's? 9's? 8's? 5's? 7's? 10's?

32. 77 are how many 10's? 9's? 8's? 7's? 6's? 11's?

33. 80 are how many 7's? 3's? 10's? 8's? 11's?

34. 96 are how many 12's? 8's? 10's? 9's? 11's?

35. 99 are how many 10's? 9's? 8's? 11's? 12's?

39. In performing each of the examples in the last three sections, we have found how many equal numbers, one of which is given, there are in another number; such a process is **Division.**

A straight line, with a dot above and below it, thus, $\div$, is the sign of division. $20 \div 5 = 4$, means 20 divided by 5 equals 4.

1. How many are 36 divided by 9? $12 \div 4$?
2. How many are 24 divided by 3? $15 \div 5$?
3. How many are $30 \div 6$, plus $22 \div 11$?
4. How many are $20 \div 2$, plus $56 \div 8$?
5. How many are $64 \div 8$, plus $84 \div 7$?
6. How many are $121 \div 11$, plus $132 \div 12$?
7. How many more are $72 \div 6$ than $45 \div 9$?
8. How many more are $63 \div 7$ than $56 \div 7$?
9. How many more are $54 \div 6$ than $90 \div 10$?
10. How many less are $110 \div 10$ than $24 \div 2$?
11. Find the difference between $49 \div 7$ and $60 \div 12$.
12. Find the difference between $88 \div 8$ and $144 \div 12$.

☞ For Slate Exercises in Division, see pages 168, 169.

Section XVI.

DIVISION, — FRACTIONAL FORM.

40. 1. If Thomas should divide 4 figs equally between his 2 brothers, how many figs would each of his brothers receive?

2. When any number is divided into 2 equal parts, what is each of the parts called?

Ans. One half of the number.

3. What is one half of 4 figs? of 8 grapes? of $10?

4. If William has 10 peaches, and gives one half of them to his brother, how many will his brother have?

5. How do you find one half of a number?

Ans. By dividing the number by 2.

6. What is one half of 12? of 14? of 16? of 18? of 2?

7. If you have 2 cents, and should spend 1 of them, what part of the 2 cents would you spend?

8. What part of 2 cents is 1 cent?

9. What part of 2 oranges is 1 orange?

10. If 6 cents are paid for 2 oranges, what part of 6 cents should be paid for 1 orange? how many cents?

11. If 6 apples are divided equally among 3 girls, how many apples will each girl have?

12. When any number is divided into 3 equal parts, what is each of the parts called?

Ans. One third of the number.

13. What is one third of 3 cents? of 12 sheep? of 15 cherries?

14. Mr. Locke had 15 bushels of potatoes, and sold one third of them; how many bushels did he sell?

15. How do you find one third of a number?

Ans. By dividing the number by 3.

16. What is one third of 18? of 21? of 27?

17. If there are 3 pitchers on the table, and you take away 1 of them, what part of the 3 pitchers do you take away? how many pitchers?

18. What part of 3 pies is 1 pie? are 2 pies?

19. If 24 lemons will make 3 gallons of lemonade, how many lemons will make 1 gallon?

20. If 4 persons share $8 equally, how many dollars will each person have?

21. When any number is divided into 4 equal parts, what is each of the parts called?

Ans. One fourth of the number.

22. What is one fourth of 4 dogs? of 8 lambs?

23. Mr. Smith earned $20 in a week: if he paid one fourth of this money for a hat, how many dollars did he pay for it?

24. How do you find one fourth of a number?

25. What is one fourth of 24? of 32? of 36?

26. If 28 pineapples are divided equally among 4 sailors, what part of the 28 pineapples will each sailor have? how many pineapples?

27. What part of 4 sailors is 1 sailor? are 2 sailors?

28. If 4 Bibles cost 48 cents, what will 1 Bible cost?

29. When any number is divided into 5 equal parts, what is each of the parts called?

Ans. One fifth of the number.

30. What is one fifth of 5 lilies? of 10 roses?

31. How do you find one fifth of a number?

32. Mr. Shaw divided 15 tulips equally among 5 girls; what part of the tulips did each girl have? how many tulips?

33. What part of 5 girls is 1 girl? are 3 girls?

34. If 5 plates cost 20 cents, what will 1 plate cost?

Solution. If 5 plates cost 20 cents, 1 plate will cost 1 fifth of 20 cents, or 4 cents; therefore, 1 plate will cost 4 cents.

35. If 5 vests cost $35, what will 1 vest cost?

36. If 5 loads of wood cost $50, what will 1 load cost?

37. When a number is divided into 6 equal parts, what is each of the parts called?

38. What is one sixth of 12 quinces? of 6 lamps?

39. How do you find one sixth of a number?

40. 24 books were divided equally among 6 pupils; how many books did each pupil receive?

41. What part of 6 pupils is 1 pupil? are 3 pupils?

42. If 6 barrels of flour cost $48, what costs 1 barrel?

43. When any number is divided into 7 equal parts, what is each of the parts called?

44. What is one seventh of 7 carriages? of 21 pens?

45. How do you find one seventh of a number?

46. If 14 acres of wheat can be mowed in 7 days, how many acres can be mowed in 1 day?

47. What part of 7 days is 1 day? are 4 days?

48. If 7 shawls cost $63, what will 1 shawl cost?

49. When any number is divided into 8 equal parts, what is each of the parts called?

50. What is one eighth of 8 bees? of 24 days?

51. How do you find one eighth of a number?

52. Mrs. Lowe divided 32 pears equally among of her scholars; how many pears did she give to each?

53. When any number is divided into 9 equal parts, what is each of the parts called?

54. What is one ninth of 18 tables? of 27 chairs?

55. How do you find one ninth of a number?

56. If 9 bags of rye cost $18, what will 1 bag cost?

57. When any number is divided into ten equal parts, what is each of the parts called?

58. What is one tenth of 10 hens? of 20 geese?

59. When any number is divided into 11 equal parts, what is each of the parts called?

60. What is one eleventh of 11? of 44? of 66?

61. When any number is divided into 12 equal parts, what is each of the parts called?

62. What is one twelfth of 36? of 48? of 72?

63. At $36 a dozen, what will be the cost of 1 fork?

64. At $10 a hundred, how many melons can be bought for $1.

65. If 1 man can do a piece of work in 66 days, in what time could 11 men do the same work?

☞ For Slate Exercises in Division, see page 170.

Section XVII.

DIVISION, — FRACTIONAL FORM.

41. 1. If 14 boys are separated into 2 equal parties, how many boys are there in each party?

2. If 5 skeins of yarn cost 55 cents, what will 1 skein cost?

3. When 4 rolls of tape can be bought for 36 cents, what is the price of 1 roll?

4. What costs 1 yard of calico, when 6 yards can be bought for 72 cents?

5. Mr. Ingalls put up 30 radishes in 5 equal bunches; how many did he put in 1 bunch?

6. If a man puts up 70 turnips in 10 equal bunches, how many turnips does he put in 1 bunch?

7. Samuel took 3 loads of apples to market, containing in all 36 bushels; how many bushels did 1 load contain?

8. If 33 prizes are distributed equally among 11 scholars, how many prizes will each scholar receive?

9. A tailor bought 10 pieces of cloth, for which he paid $110; what did he pay for each piece?

10. If 7 sheep can be bought for $56, what will 1 sheep cost?

11. If 8 spools of cotton can be bought for 96 cents, what will 1 spool cost?

12. If 9 men receive $63 as prize money, how many dollars does 1 man receive?

13. If 11 yards of cloth cost 132 cents, how many cents does 1 yard cost?

14. What is the price of 1 candle, when 12 candles can be bought for 108 cents?

15. If 1 man can set the type for a book in 44 days, in how many days can 4 men do the same work?

16. If 1 man can earn a certain sum of money in 42 days, in how many days can 6 men earn the same sum?

17. A cistern has 2 equal pipes; if by 1 pipe the cistern could be emptied in 24 minutes, in how many minutes could it be emptied by both pipes?

18. How many days would it take 7 men to earn as much as 1 man can earn in 49 days?

19. If 60 men can do a piece of work in 1 day, how many men could do the same work in 5 days?

20. If a quantity of provision will last 28 persons 1 day, how many persons would the same provision last 4 days?

21. If 27 horses eat a ton of hay in 1 week, how many horses would eat a ton in 3 weeks?

22. Charles had 24 apples, and gave 1 third of them to his sister, and 1 sixth to a schoolmate; how many apples did he give to his sister? how many did he give to his schoolmate? how many to both? how many apples did he keep himself?

23. 1 sixth of 18, plus 1 half of 22, are how many?

24. 1 fifth of 35, plus 1 sixth of 24, are how many?

25. 1 third of 36, minus 1 ninth of 45, are how many?

26. 1 tenth of 80, minus 1 fifth of 40, are how many?

27. 1 third of 33, plus 1 eighth of 56, are how many?

28. 1 half of 26, minus 1 third of 15, are how many?

29. 1 tenth of 110, plus 1 fifth of 60, are how many?

30. 1 twelfth of 144, minus 1 ninth of 36, are how many times 4 (or how many 4's)?

42. 1. If 3 boxes of currants cost 24 cents, what will 2 boxes cost?

Solution. If 3 boxes cost 24 cents, 1 box will cost 1 third of 24 cents, which is 8 cents, and 2 boxes will cost 2 times 8 cents (or two 8's of cents), which equals 16 cents; therefore, 2 boxes will cost 16 cents.

2. If 4 baskets can be bought for 36 cents, what will 1 basket cost? what will 3 baskets cost?

3. What is 1 fourth of 36? 3 fourths of 36?

4. If 5 barrels of flour cost $45, what will 4 barrels cost?

5. What is 1 fifth of 45? 4 fifths of 45?

6. If 6 yards of muslin cost 60 cents, what will 5 yards cost?

7. What is 1 sixth of 60? 5 sixths of 60?

8. If a horse can travel 32 miles in 4 hours, how far can he travel in 3 hours?

9. What is 1 fourth of 32? 3 fourths of 32?

10. A man rode 77 miles in 11 hours; how far, at the same rate, would he ride in 4 hours?

11. What is 1 eleventh of 77? 4 elevenths of 77?

12. A man hired a laborer for 9 weeks, and paid him $81; what did he pay him for 4 weeks' labor?

13. If 84 panes of glass are required for 7 windows, how many panes will be required for 12 windows?

14. If there are 72 windows in 6 cottages, how many windows are there in 7 cottages?

15. What will 4 oranges cost, at the rate of 3 oranges for 12 cents?

16. A farmer exchanged 5 sheep for fowls, at the rate of 2 sheep for 20 fowls; how many fowls did he receive?

17. At the rate of 2 melons for 24 peaches, how many peaches can be bought for 8 melons? for 9 melons?

18. How many are 3 fifths of 25? of 60?

19. How many are 5 sevenths of 63? of 77?

20. How many are 7 ninths of 81? of 108?

21. How many are 9 tenths of 90? of 110?

22. How many are 11 twelfths of 132? of 144?

☞ For Slate Exercises in Division, see page 171.

Section XVIII.

EXAMPLES IN MULTIPLICATION AND THE FRACTIONAL FORM OF DIVISION.

43. 1. Charles sold 3 dozen eggs at 10 cents a dozen, and with the money that he received for them, he bought 5 pounds of rice; how many cents did 1 pound of rice cost?

2. What is 1 fifth of 3 times 10 (or three 10's)?

3. Horace bought 2 cocoanuts at 12 cents apiece, and gave in payment 4 heads of lettuce; what was the lettuce worth a head?

4. What is 1 fourth of 2 times 12 (or two 12's)?

5. Mr. Allen took 8 steps, moving onward a distance of 3 feet at each step: if Henry took 12 steps in going the same distance, how far did Henry go at each step?

6. What is 1 twelfth of 8 times 3 (or eight 3's)?

7. Mary's middle finger is 4 inches long, and it takes 9 times the length of her finger to measure a yard; how many inches are there in a yard? Etta has to take 12 times the length of her middle finger to measure a yard; how many inches long is Etta's middle finger?

8. What is 1 twelfth of 9 times 4 (or nine 4's)?

9. If 2 men can dig a ditch in 6 days, in how many days can 1 man dig it? In how many days can 4 men dig it?

10. If a quilt can be made by 3 girls in 8 days, in how many days can it be made by 12 girls?

11. If by working 6 hours a day, a piece of work can be done in 2 days, in how many days can it be done by working 1 hour a day? 3 hours? 4 hours?

12. If a man can perform a journey in 8 hours by travelling 6 miles an hour, in how many hours can he perform it by travelling 4 miles an hour?

13. If a firkin of butter will last a family of 6 persons 10 weeks, how many weeks would the same butter last 5 persons?

14. If a lot of grain will last 12 horses 4 days, how many horses will it last 6 days?

15. How many men will be required to hew as much timber in 12 days as 8 men can hew in 9 days?

16. What is 1 third of 9 times 4 (or nine 4's)?

17. What is 1 sixth of 9 times 8 (or nine 8's)?

18. What is 1 eighth of 4 times 12?

19. What is 1 ninth of 6 times 6?

20. What is 1 tenth of 8 times 5?

21. What is 1 twelfth of 6 times 8?

22. What is 1 eleventh of 2 times 22?

☞ For Slate Exercises in Division, see page 172.

Section XIX.

TO FIND THE WHOLE NUMBER WHEN A PART IS GIVEN.

44. 1. Some pictures were divided equally between Albert and Warren; what part of the pictures did each boy receive? If one of the boys received 3 pictures, how many pictures did both of the boys receive?

2. 3 is 1 half of what number?

Solution. If 3 is 1 half of some number, 2 halves, or the whole number, must be 2 times 3, which are 6; therefore, 3 is 1 half of 6.

3. 5 is 1 half of what number?

4. 6 is 1 half of what number?

5. How many halves of a number equal the whole number?

6. When 1 half of a number is given, how do you find the whole number?

7. Mary gave 3 pears to Lucy, which was 1 half of all she had; how many had she?

8. Three girls divided some roses equally among themselves; what part of the roses did each girl have? If each girl had 2 roses, how many had all the girls?

9. 2 is 1 third of what number?

10. 7 is 1 third of what number?

11. How many thirds of a number equal the whole number?

12. When 1 third of a number is given, how do you find the whole number?

13. Sarah spent 6 cents for candy, which was 1 third of all the money she had; how many cents had she?

14. Four boys divided some books equally among themselves; what part of the books did each boy have? If each boy had 2 books, how many books did all the boys have?

15. 2 is 1 fourth of what number?

16. 9 is 1 fourth of what number?

17. How many fourths of a number equal the whole number?

18. When 1 fourth of a number is given, how do you find the whole number?

19. Felix missed 6 words in spelling, which was 1 fourth of all the words missed by the whole school; how many words were missed by the whole school?

20. Agnes is 3 years old, which is 1 fifth of the age of her brother; what is the age of her brother?

21. How many fifths of a number equal the whole number?

22. Mr. Gray is hired to gather apples for Mr. Frost, and is to keep 1 bushel of apples for every 6 bushels gathered; what part of the apples gathered does Mr. Gray keep? If Mr. Gray keeps 4 bushels, how many bushels are gathered?

23. How many sixths of a number equal the whole number?

24. Mr. Jones pays $3 for a vest, which is 1 seventh as much as he pays for a coat; how much does he pay for a coat?

25. How many sevenths of a number equal the whole number?

26. Mr. Tyler is an auctioneer who sells goods and keeps 1 eighth of the money received, to pay him for selling; if he keeps $5, how many dollars' worth are sold?

27. How many eighths of a number equal the whole number?

28. Alvin gave 6 cents for a pencil, which was 1 ninth as much as he gave for a book; how many cents did he give for the book?

29. How many ninths of a number equal the whole number?

30. How many tenths of a number equal the whole number? how many elevenths? how many twelfths?

31. Edith has spent 2 weeks at a boarding-school, which is 1 eleventh of the term; how many weeks are there in the term?

32. After travelling 11 miles, Mr. Ives finds he has performed 1 tenth of his journey; how many miles had he to travel?

33. Sarah has learned 8 pages of her Arithmetic, which is 1 twelfth of the whole book; how many pages are there in the whole book?

34. 11 is 1 sixth of what number?

35. 10 is 1 twelfth of what number?

36. 25 is 1 half of what number?

37. 15 is 1 third of what number?

38. 5 is 1 twentieth of what number?

39. 30 is 1 fourth of what number?

☞ For Slate Exercises, perform examples on pages 171 and 172, changing the left hand figure of each divisor to a figure denoting one more.

Section XX.

General Review.

45. 1. What is a unit?
2. What is a number?
3. What is addition?
4. Give an example in addition.
5. What is subtraction?
6. Give an example in subtraction.
7. What is multiplication?
8. Give an example in multiplication.
9. What is division?
10. Give an example in division.

11. Loring found 17 pears under a tree, and gave 2 to Mary, 3 to John, and 7 to his mother; how many did he give away? How many had he left?

12. What number must be put with 18 to make 30?

13. What number must be taken out of 75 to leave 62?

14. The sum of two numbers is 48, and one of the numbers is 16; what is the other number?

15. The sum of three numbers is 50: two of the numbers are 12 and 13; what is the third number?

16. Clarence is 3 years younger than Arthur, and 4 years older than George: if Clarence is 12 years old, how old is Arthur? How old is George?

17. I bought at a confectioner's some candy for 15 cents, some oranges for 25 cents, some nuts for 18 cents, and some raisins for 20 cents, and gave $1 in payment; what change should I receive?

18. A man bought 5 calves at $8 apiece, and gave in payment 3 barrels of mackerel at $12 a barrel, and paid the balance in money; how many dollars in money did he pay?

19. A merchant bought 8 sheep at $6 apiece, and paid as much of the cost as possible in 10-dollar bills, as much of the balance as possible in 5-dollar bills, and the rest in 1-dollar bills; how many bills of each sort did he pay?

20. Mr. Williams had 2 boxes of hats, with 10 hats in each box, and 4 hats besides; Mr. Frank had 3 boxes of hats, with 11 hats in each box, and 9 hats besides; how many more hats had Mr. Frank than Mr. Williams?

21. A man exchanged 7 pounds of butter for cheese, at the rate of 10 pounds of butter for 30 pounds of cheese; how many pounds of cheese did he receive?

22. If 1 man can build a cistern in 12 days, how many men will be required to build it in 1 day?

23. If it costs $48 for bricks to build a cistern, when bricks are worth $8 a thousand, what will it cost for bricks to build it when bricks are worth $10 a thousand?

24. When apples are worth $5 a barrel, and coal $10 a ton, how many barrels of apples will pay for 6 tons of coal?

25. If a number of 5-cent loaves weigh 10 pounds, what will the same number of 12-cent loaves weigh?

26. A man raised 36 bushels of potatoes, which was 12 times as many as he planted; how many bushels did he plant?

27. Ezra had 48 plums; if he gave 1 sixth of them to his sister, and 3 eighths to his brother, how many plums had he left?

28. A man paid $4 for a week's board, which was 1 sixth of what he earned; how much did he earn?

29. Take 5, multiply it by 4, add 5, divide by 5, subtract 1, subtract 2, multiply by 12, add 3, take 1 ninth of that number, divide by 3, and subtract 1; what number have you?

☞ For Slate Exercises, see page 173.

Section XXI.

MULTIPLES AND FACTORS.

46. 1. Name some of the multiples of 2; of 3.

The *multiples* of 2 are . . 2, 4, **6**, 8, 10, **12**, etc.
The *multiples* of 3 are . . 3 **6**, 9, **12**, etc.

NOTE. 6 and 12 are multiples of both 2 and 3; hence 6 and 12 are **Common Multiples** of 2 and 3; and 6 is the **Least Common Multiple** of 2 and 3.

2. What numbers from 3 to 24 are multiples of 3? of 4? of both 3 and 4?

3. What is the least common multiple of 3 and 4? of 4 and 6? of 9 and 6? of 5 and 6? of 12 and 9?

4. What is a multiple of a number? *Ans.* Any number which will contain it without a remainder.

5. What is a common multiple of two or more numbers? the least common multiple?

47. 1. Of what numbers is 6 a multiple? *Ans.* 6 is a multiple of 2 and 3.

NOTE. 2 and 3 are called the **Factors** of 6.

2. What are the factors of 18?

Ans. 3 with 6; 2 with 9; 2 with 3 with 3.

3. What are the factors of 20? 22? 24? 25? 36?

4. What are the factors of 2? of 7? *Ans.* They have no factors besides themselves and 1.

Numbers which have no factors besides themselves and 1 are called **Prime** numbers.

Numbers which have other factors besides themselves and 1 are called **Composite** numbers.

5. Is 3 a composite number? is 4? is 5? is 6?

6. Name the prime numbers from 2 to 30.

7. Name the composite numbers from 2 to 50.

Section XXII.

48. FRACTIONS.

When an orange is divided into 2 equal parts, each of the parts is one half of an orange.

1. Alvin cut an orange into two equal parts, and gave one of the parts to his brother. What part of the orange did each then have?

2. When anything is divided into 2 equal parts, what is each of the parts called?

3. How many halves equal a whole thing?

When an orange is divided into 3 equal parts, each of the parts is one third of an orange.

4. Three boys divided a melon equally among themselves; what part of the melon did each boy have?

5. When anything is divided into 3 equal parts, what is each of the parts called? What are 2 of the parts called?

6. How many thirds equal a whole thing?

7. Mr. Stone cut a cheese into 3 equal parts: he kept one of the parts, and gave the rest to his neighbors; what part of the cheese did he keep? How many thirds did he give to his neighbors?

Fourths.

Fifths.

Sixths.

8. When anything is divided into 4 equal parts, what is each of the parts called? What are 2 of the parts called? 3 of the parts?

9. When anything is divided into 5 equal parts, what is each of the parts called? What are 2 of the parts called? 3 of the parts? 4 of the parts?

10. When anything is divided into 6 equal parts, what is each of the parts called? What are 2 of the parts called? 3 of the parts? 4 of the parts?

11. How many fourths equal a whole thing? how many fifths? how many sixths?

12. How many sevenths equal a whole thing? how many ninths? how many tenths? elevenths? twelfths?

13. Which is the larger, 1 half or 1 third of anything? 1 third or 1 fourth? 1 fifth or 1 fourth?

14. Mrs. Day gave 1 fifth of a melon to Jane, and 2 fifths to Mary; how many fifths did she give to both? How many fifths remained?

15. How many fifths remain, if you take 1 fifth and 2 fifths out of 5 fifths?

16. How many sixths are 2 sixths and 3 sixths? 8 ninths and 4 ninths?

17. How many tenths are 7 tenths less 3 tenths? 7 twelfths less 5 twelfths?

18. If 1 pineapple costs 1 fifth of a dollar, what will 2 pineapples cost?

19. Ira gave to 4 persons 1 sixth of a barrel of flour apiece; how many sixths did he give to all?

20. At the rate of 1 eighth of a dollar an hour, what will a woman charge for 2 hours' work? for 3 hours' work? for 5 hours' work?

21. If 2 sevenths of a candle are consumed in 1 hour, in how many hours will 4 sevenths be consumed?

22. 6 sevenths are how many 2's of sevenths?

23. 8 ninths are how many 2's of ninths? 4's of ninths?

24. Mr. Brown divided 6 eighths of a bar of silver into 3 equal parts; how many eighths were there in each part?

25. What is 1 third of 6 eighths? 1 half of 6 eighths?

26. What is meant by 1 half of a thing? *Ans.* One of the two equal parts into which the thing is divided.

27. What is meant by 1 third of a thing?

28. What is meant by 2 thirds of a thing? *Ans.* Two of the three equal parts into which the thing is divided.

29. What is meant by 1 fourth of a thing? by 2 fourths? 3 fourths?

30. Into how many equal parts must anything be divided to make fifths? sixths? sevenths? eighths? tenths? twelfths? thirteenths?

Section XXIII.

FRACTIONS, — REDUCTION.

49. 1. If you should cut each of 3 apples into halves, how many halves would there be?

2. How many halves are there in 3? in 4? in 5?

3. I have 3 and 1 half sheets of paper; how many letters can I write upon this paper, if I use 1 half of a sheet for each letter?

4. How many halves are there in 3 and 1 half?

Solution. If in 1 there are 2 halves, in 3 there are 3 times 2 halves (or three 2's of halves), which equals 6 halves; 6 halves and 1 half are 7 halves; therefore, in 3 and $\frac{1}{2}$ there are 7 halves.

5. How many halves are there in 4 and 1 half? how many halves in 5 and 1 half? how many halves in 6 and 1 half?

6. How many halves are there in any number?

Ans. The number of halves in any number equals the number of units multiplied by 2.

7. A man had 2 bushels of oats, and wished to give 1 third of a bushel apiece to some horses; to how many horses could he give them?

8. How many thirds are there in 4? in 6?

9. Annie puts 1 third of a cup of butter into each cake; how many cakes can she make if she uses 2 and 1 third cups of butter?

Solution. If she puts 1 third of a cup of butter into 1 cake, with 2 and 1 third cups she can make as many cakes as there are thirds in 2 and 1 third. In 1 there are 3 thirds, etc.

10. If John earns 1 third of a dollar in 1 day, in how many days can he earn 4 and 1 third dollars?

11. How many thirds are there in 4 and 1 third? in 5 and 2 thirds?

12. How many thirds are there in any number?

13. If it requires 1 fourth of a yard of cloth to make 1 cap, how many caps can be made out of 5 yards? out of 5 and 1 fourth yards?

14. How many fourths are there in 5? in 10 and 3 fourths?

15. How many fourths are there in any number?

16. If a man earns 1 fifth of a dollar in an hour, in how many hours can he earn 4 dollars? 4 and 2 fifths dollars? 5 and 4 fifths dollars?

17. How many fifths are there in 4? in 5 and 1 fifth?

18. How many fifths are there in any number?

19. Mrs. Clarke gave 1 sixth of a pie to each member of her family, which took 1 pie and 4 sixths of a pie; how many persons were there in her family?

20. How many sixths are there in 4 and 1 sixth? in 5 and 2 sixths?

21. How many sixths are there in any number?

22. If Emma can do 1 seventh of her task in 1 hour, in how many hours can she perform her whole task?

23. How many sevenths are there in 1? in 2 and 5 sevenths? in 4 and 4 sevenths?

24. How many sevenths are there in any number?

25. Mr. G. divided 2 acres of land into house-lots of 1 eighth of an acre each; how many house-lots had he?

26. How many eighths are there in 2? in 2 and 5 eighths? in 8 and 7 eighths?

27. How many eighths are there in any number?

28. How many ninths are there in 7? in 8 and 2 ninths? in 10 and 4 ninths?

29. How many tenths are there in 6 and 3 tenths? in 7 and 6 tenths? in 11 and 3 tenths?

30. How many elevenths are there in 4 and 7 elevenths? in 7 and 10 elevenths?

31. How many twelfths are there in 8 and 9 twelfths?

32. How many ninths are there in any number? how many tenths? elevenths? twelfths?

50. 1. If you give 2 girls 1 half of a sheet of paper apiece, how many sheets does it require?

2. How many ones, or units, are there in 2 halves? in 4 halves?

3. How many units are there in 5 halves?

Solution. — Since there are 2 halves in 1 unit, in 5 halves there are as many units as there are 2's in 5; there are two 2's in 5, and 1 remains; therefore, in 5 halves there are 2 units and 1 half.

4. How many units are there in 3 halves? in 9 halves? in 10 halves? in 11 halves? in 13 halves?

5. How many units are there in any number of halves?

6. If a family consume 1 third of a barrel of flour in 1 week, how many thirds will they consume in 6 weeks? how many barrels?

Solution. If they consume 1 third of a barrel in 1 week, in 6 weeks they will consume 6 times 1 third of a barrel, or 6 thirds, which equals 2 barrels; therefore, they will consume 2 barrels.

7. If I give to each of 4 children 1 third of a cake, how many thirds of a cake do I give to all? how many cakes?

8. If 8 boys earn 1 third of a dollar apiece in a day, how many dollars will all earn?

9. How many units are there in 7 thirds? in 8 thirds? in 9 thirds? in 11 thirds? in 13 thirds? in 17 thirds?

10. How many units are there in any number of thirds?

11. At 1 fourth of a dollar apiece, what cost 4 trees? 5 trees? 6 trees?

12. How many units are there in 4 fourths? in 5 fourths? in 7 fourths? in 10 fourths? in 15 fourths?

13. How many units are there in any number of fourths?

14. If 1 fifth of a pound of soap will pay for 1 bunch of matches, how many pounds of soap will pay for 6 bunches of matches?

15. How many units are there in 10 fifths? in 12 fifths? in 14 fifths? in 17 fifths?

16. How many units are there in any number of fifths? of sixths? of eighths? of tenths? of sevenths? of ninths? of twelfths?

17. At 1 sixth of a dollar apiece, how many dollars will pay for 6 baskets? for 8 baskets?

18. If 1 seventh of a candle will be consumed in 1 evening, how many candles will last 10 evenings?

19. If in 1 house-lot there is 1 eighth of an acre of land, how many acres will be required to make 20 house-lots? to make 25 house-lots?

20. How many units are there in 18 sixths? in 22 sixths? in 18 sevenths? in 19 eighths? in 16 ninths? in 23 tenths? in 17 elevenths? in 33 twelfths?

Section XXIV.

Definitions and General Principles.

51. In the last two sections the pupil has been taught to operate upon parts of units; one of the equal parts of a unit is a **Fraction,** or a **Fractional Unit.** A collection of fractional units is a **Fractional Number.**

The following is the common method of expressing fractions and fractional numbers by figures:

1 half is expressed thus, $\frac{1}{2}$	2 thirds is express'd thus, $\frac{2}{3}$
1 third " " $\frac{1}{3}$	3 fourths " $\frac{3}{4}$
1 fourth " " $\frac{1}{4}$	5 sixths " $\frac{5}{6}$
1 sixth " " $\frac{1}{6}$	2 and 7 eighths " $2\frac{7}{8}$
1 eleventh " " $\frac{1}{11}$	5 and 4 fifteenths " $5\frac{4}{15}$

1. Express in figures 1 third; 1 fifth; 1 seventh; 1 tenth; 2 thirds; 2 fourths; 4 sevenths; 3 and 1 ninth.

2. How many numbers are used in expressing each of these fractional numbers? what else is used?

3. What is the number expressed below the line.

Ans. It is the number of equal parts into which a unit is divided.

4. What name is given to this number?

Ans. The **Denominator** of the fraction or fractional number; because it gives a *name* or *denomination* to the fractional parts; thus, if the denominator is 3, the fractional parts are thirds; if 4, fourths, etc.

5. Name the denominator of each of the following: $\frac{1}{2}$; $\frac{2}{9}$; $\frac{3}{12}$; $\frac{7}{13}$; $\frac{9}{14}$; $\frac{5}{22}$.

6. What does the denominator of the fraction $\frac{1}{2}$ show? of $\frac{2}{9}$?

7. What is the number expressed above the line?

Ans. It is the number of equal parts of the unit which are taken.

8. What name is given to this number?

Ans. The **Numerator** of the fraction or fractional number; because it *numbers* the fractional parts taken.

9. Name the numerator of each of the following: $\frac{2}{5}$; $\frac{1}{8}$; $\frac{5}{6}$; $\frac{7}{4}$; $\frac{8}{4}$.

The numerator and denominator are called the **Terms** of the fraction or fractional number.

In the fractional numbers $\frac{7}{4}$ and $\frac{8}{4}$, expressed above, there are more fourths than are found in 1 unit, $\frac{7}{4}$ being equal to $1\frac{3}{4}$, and $\frac{8}{4}$ being equal to 2. Such fractional numbers can always be changed to a whole number, or to a whole with a fractional number. A number, as $1\frac{3}{4}$ above, consisting of a whole number with a fractional number, is a **Mixed Number.**

1. Change $\frac{5}{4}$ to a mixed number.
2. Change $\frac{10}{3}$ to a mixed number.
3. Change $\frac{84}{7}$ to a whole number.

52. 1. John and Thomas each had an orange, and each gave Mary $\frac{1}{3}$ of his orange; how many thirds of an orange had Mary? How many thirds of an orange had each of the boys left?

2. Is there any difference between $\frac{1}{3}$ of 2 oranges and $\frac{2}{3}$ of 1 orange? Then, $\frac{1}{3}$ of 2 equals what part of 1?

By the following diagrams it will be seen that $\frac{1}{4}$ of 3 is equal in value to $\frac{3}{4}$ of 1.

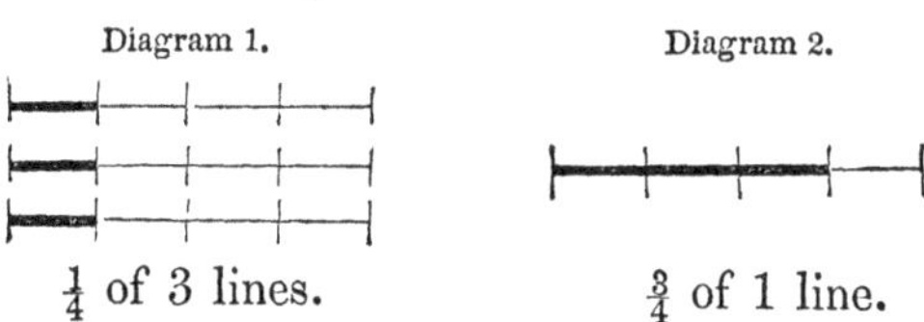

$\frac{1}{4}$ of 3 lines. $\frac{3}{4}$ of 1 line.

The pupil may show, in the same way, that $\frac{1}{4}$ of 2 is equal in value to $\frac{2}{4}$ of 1; that $\frac{1}{5}$ of 2 equals $\frac{2}{5}$ of 1.

3. What is meant by $\frac{3}{5}$? *Ans.* 3 of the 5 equal parts into which 1 is divided, or $\frac{1}{5}$ of 3.

4. What is meant by $\frac{5}{6}$? $\frac{2}{7}$? $\frac{8}{11}$? $\frac{9}{10}$? $\frac{7}{12}$?

53.

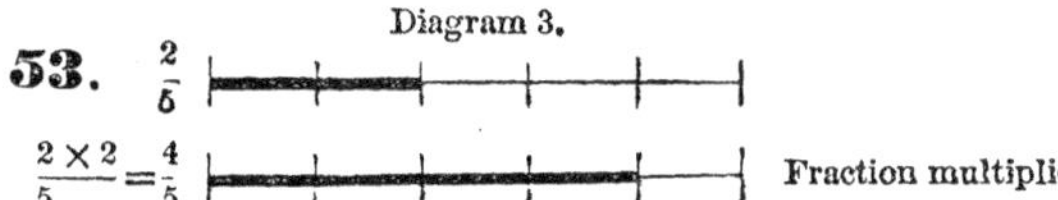

I. Diagram 3 shows that if we multiply the numerator of a fraction, we multiply the fraction; for we multiply the number of parts, while their size remains the same.

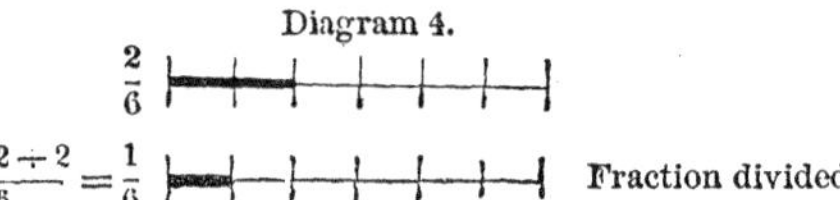

II. Diagram 4 shows that if we divide the numerator of a fraction, we divide the fraction; for we divide the number of parts, while their size remains the same.

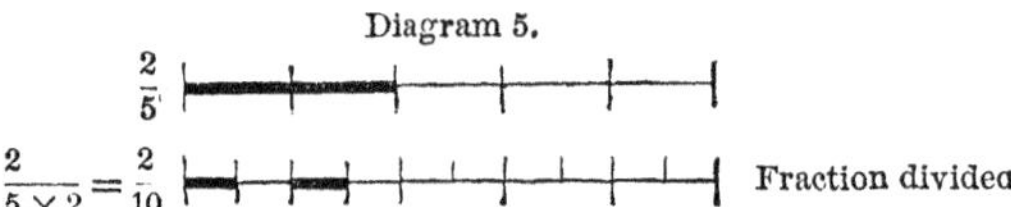

III. Diagram 5 shows that if we multiply the denominator of a fraction, we divide the fraction; for we diminish the size of the parts, while the number of parts remains the same.

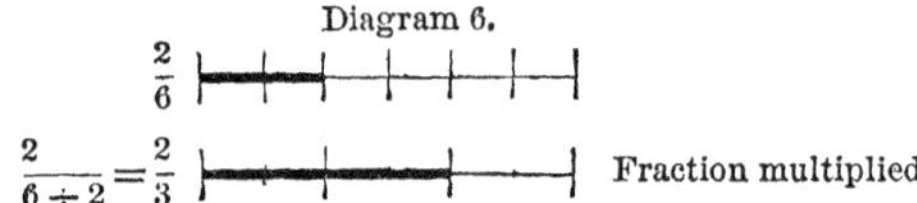

IV. Diagram 6 shows that if we divide the denominator of a fraction, we multiply the fraction; for we increase the size of the parts, while the number remains the same.

1. How does *multiplying the numerator* of a fraction affect the value of the fraction? Why?

2. How does *dividing the numerator* affect the value of the fraction? Why?

3. How does *multiplying the denominator* of a fraction affect the value of the fraction? Why?

4. How does *dividing the denominator* affect the value of the fraction? Why?

5. In what two ways, then, can you multiply a fraction?

6. In what two ways can you divide a fraction?

7. Multiply $\frac{2}{5}$ by 2; $\frac{3}{4}$ by 2; $\frac{4}{5}$ by 3; $\frac{5}{6}$ by 3.

8. Divide $\frac{2}{5}$ by 2; $\frac{3}{4}$ by 2; $\frac{4}{7}$ by 4; $\frac{2}{3}$ by 4.

Diagram 7.

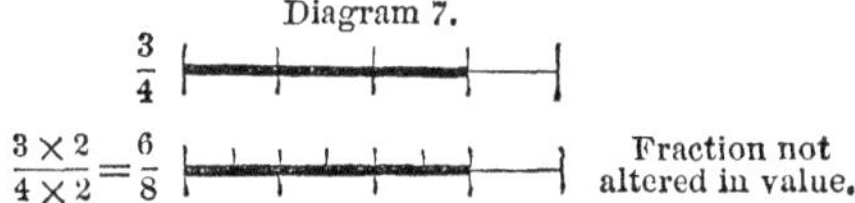

V. Diagram 7 shows that if we multiply the numerator and denominator of a fraction each by the same number, the value of the fraction is not altered. Why?

Diagram 8.

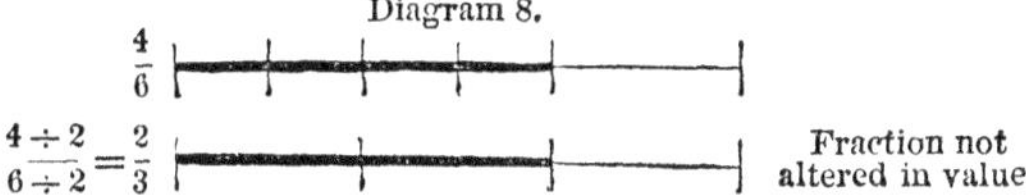

VI. Diagram 8 shows that if we divide the numerator and denominator of a fraction, each by the same number, the value of the fraction is not altered. Why?

Section XXV.

REDUCTION OF FRACTIONS.

54. 1. If an apple is divided into halves, and one of the halves is divided into 2 equal parts, what part of the apple is 1 of these pieces? What part of the apple are the 2 pieces?

2. Is there any difference in value between $\frac{2}{4}$ and $\frac{1}{2}$?

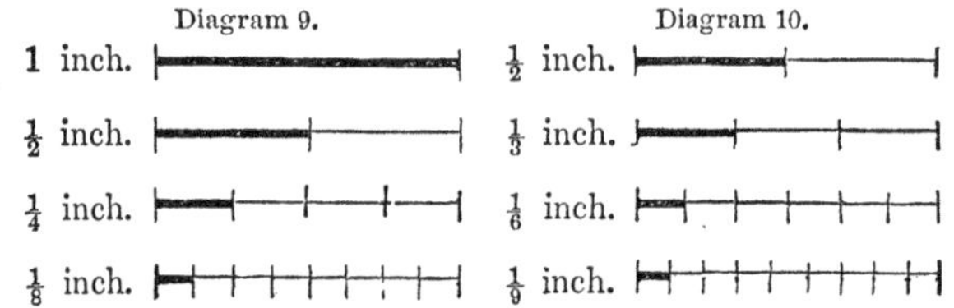

The pupil may answer the following questions by referring to the diagrams above.

3. $\frac{1}{2}$ equals how many fourths? how many eighths?
4. $\frac{1}{4}$ equals how many eighths?
5. $\frac{1}{2}$ equals how many sixths?
6. $\frac{1}{3}$ equals how many sixths? how many ninths?

7. $\frac{1}{5}$ equals how many tenths? *Solution.* Since 1 whole one equals 10 tenths, $\frac{1}{5}$ must equal $\frac{1}{5}$ of 10 tenths, which is 2 tenths; therefore, $\frac{1}{5}$ equals $\frac{2}{10}$.

NOTE. The same result may be derived from Proposition V., page 73; for as both terms of the fractional number are multiplied by the same number, 2, the value of the fractional number remains the same.

8. Change $\frac{1}{6}$ to twelfths; $\frac{1}{4}$ to twelfths.
9. Change $\frac{1}{7}$ to fourteenths; to twenty-firsts.
10. Change $\frac{1}{5}$ to fifteenths; $\frac{2}{5}$ to fifteenths.

Solution. $\frac{1}{5}$ equals $\frac{3}{15}$, and $\frac{2}{5}$ will equal 2 times $\frac{3}{15}$ (or two 3's of 15ths), which equals $\frac{6}{15}$; therefore, etc.

11. Change $\frac{2}{3}$ to ninths; $\frac{2}{5}$ to tenths; $\frac{5}{6}$ to twelfths.

55. The fractional numbers $\frac{9}{12}$ and $\frac{10}{12}$ have their denominators alike; fractional numbers whose denominators are alike have a **Common Denominator.**

1. Change $\frac{1}{2}$ and $\frac{1}{3}$ to fractional numbers having a common denominator.

NOTE. The common denominator of these fractional numbers may be either 6, 12, 18, or any common multiple of the denominators, 2 and 3; but for convenience we should select the *least* common multiple (Art. 47): 6 is the least common multiple of 2 and 3; therefore, $\frac{1}{2}$ and $\frac{1}{3}$ should be changed to 6ths.

Change the following to fractional numbers having a common denominator:

2. $\frac{1}{2}$ and $\frac{1}{5}$; $\frac{1}{3}$ and $\frac{1}{4}$; $\frac{1}{2}$ and $\frac{2}{7}$.
3. $\frac{4}{5}$ and $\frac{3}{7}$; $\frac{5}{6}$ and $\frac{2}{3}$; $\frac{2}{3}$ and $\frac{2}{9}$.
4. $\frac{4}{5}$ and $\frac{3}{4}$; $\frac{5}{6}$ and $\frac{2}{9}$; $\frac{4}{5}$ and $\frac{5}{6}$.
5. $\frac{2}{3}$ and $\frac{6}{7}$; $\frac{3}{8}$ and $\frac{7}{9}$; $\frac{4}{5}$, $\frac{5}{8}$, and $\frac{1}{4}$.
6. $\frac{7}{9}$ and $\frac{5}{7}$; $\frac{2}{7}$ and $\frac{5}{6}$; $\frac{1}{2}$, $\frac{1}{3}$, and $\frac{1}{5}$.
7. $\frac{1}{6}$ and $\frac{5}{8}$; $\frac{7}{6}$ and $\frac{4}{7}$; $\frac{3}{4}$, $\frac{5}{3}$, $\frac{1}{6}$, and $\frac{7}{12}$.

56. Arthur had 2 fourths of a dollar, which he exchanged for 1 piece of money equal in value; what part of a dollar was the piece he received?

Diagram 11.	Diagram 12.
1 inch.	$\frac{1}{2}$ inch.
$\frac{1}{2}$ inch.	$\frac{1}{3}$ inch.
$\frac{1}{4}$ inch.	$\frac{1}{6}$ inch.
$\frac{1}{8}$ inch.	$\frac{1}{9}$ inch.

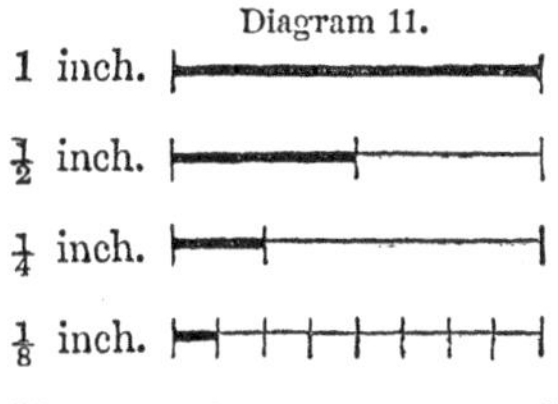

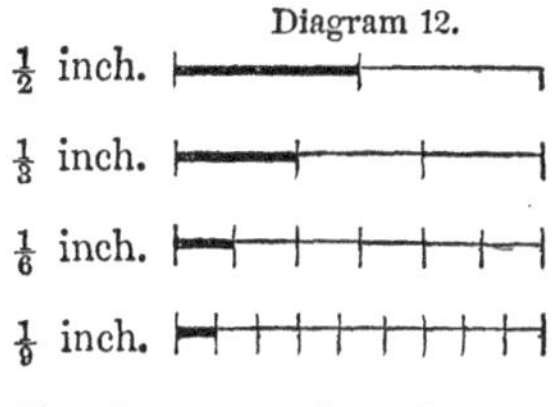

The pupil may answer the following questions by referring to the diagrams above.

1. How many halves are there in $\frac{2}{4}$? in $\frac{4}{8}$?
2. How many fourths are there in $\frac{2}{8}$? in $\frac{4}{8}$? $\frac{6}{8}$?
3. How many halves are there in $\frac{3}{6}$?
4. How many thirds are there in $\frac{2}{6}$? in $\frac{4}{6}$? $\frac{3}{9}$? $\frac{6}{9}$?

5. How many fifths are there in $\frac{8}{10}$?

Solution. 1 equals $\frac{10}{10}$; $\frac{1}{5}$ of 1 equals $\frac{1}{5}$ of $\frac{10}{10}$, or $\frac{2}{10}$. Since $\frac{1}{5}$ equals $\frac{2}{10}$, $\frac{8}{10}$ must equal as many fifths as there are 2's in 8; there are four 2's in 8; therefore, $\frac{8}{10}$ equal $\frac{4}{5}$.

6. How many 5ths are there in $\frac{6}{10}$? in $\frac{12}{15}$?
7. How many 7ths are there in $\frac{10}{14}$? in $\frac{6}{21}$?
8. How many 8ths are there in $\frac{4}{16}$? in $\frac{9}{24}$?
9. How many 4ths are there in $\frac{15}{20}$? in $\frac{21}{28}$?
10. How many 4ths are there in $\frac{6}{8}$? in $\frac{36}{48}$?
11. How many 9ths are there in $\frac{16}{18}$? in $\frac{8}{36}$?
12. How many 11ths are there in $\frac{6}{22}$? in $\frac{12}{33}$?

57. A fractional number is changed to its smallest terms when it is expressed by the smallest numbers; that is, when there is no number greater than 1 that will divide both numerator and denominator without a remainder. [See Prop. VI., page 73.]

1. Change $\frac{4}{6}$ to its smallest terms.

NOTE. Divide numerator and denominator by 2. *Ans.* $\frac{2}{3}$.

2. Change $\frac{6}{8}$ to its smallest terms. Change $\frac{8}{12}$.

NOTE. $\frac{8 \div 4 = 2}{12 \div 4 = 3}$; or $\frac{8 \div 2 = 4}{12 \div 2 = 6}$, $\frac{4 \div 2 = 2}{6 \div 2 = 3}$.

3. Change to their smallest terms $\frac{2}{10}$; $\frac{5}{10}$; $\frac{3}{9}$.
4. Change to their smallest terms $\frac{8}{20}$; $\frac{12}{20}$; $\frac{4}{22}$.
5. Change to their smallest terms $\frac{4}{24}$; $\frac{8}{32}$; $\frac{14}{16}$.
6. Change to their smallest terms $\frac{9}{27}$; $\frac{9}{72}$; $\frac{15}{35}$.
7. Change to their smallest terms $\frac{40}{45}$; $\frac{55}{60}$; $\frac{18}{36}$.
8. Change to their smallest terms $\frac{27}{63}$; $\frac{42}{72}$; $\frac{72}{99}$.
9. Change to their smallest terms $\frac{28}{56}$; $\frac{14}{42}$; $\frac{88}{110}$.
10. Change to their smallest terms $\frac{8}{22}$; $\frac{20}{16}$; $\frac{32}{40}$.
11. Change to their smallest terms $\frac{24}{48}$; $\frac{24}{144}$; $\frac{121}{55}$.

Section XXVI.

ADDITION AND SUBTRACTION OF FRACTIONS.

58. 1. A grocer sold a pound of butter for 2 fifths of a dollar, and a pound of cheese for 1 fifth of a dollar; how much did he receive for both?

2. I bought a mat for 8 tenths of a dollar, and a pail for 3 tenths; how much did I pay for both?

3. Helen had a large orange, but she has given away 2 fifths of it; what part of the orange has she left?

4. 1 less 2 fifths equals how many fifths?

5. Jane bought a Dictionary for $\$\frac{7}{8}$, a slate for $\$\frac{3}{8}$, and an Arithmetic for $\$\frac{6}{8}$; how much did she pay for all?

6. Charles has picked $\frac{5}{9}$ of a peck of berries, William has picked $\frac{8}{9}$ of a peck, and Alfred $\frac{7}{9}$; how many pecks have they all picked?

7. If a pineapple cost $2\frac{2}{5}$ dimes, and a lobster cost $1\frac{2}{5}$ dimes, what did both cost?

8. How many are $4\frac{4}{5}$ plus $3\frac{3}{5}$? $1\frac{3}{11}$ plus $2\frac{10}{11}$?

9. Mr. Blake owed William $\$\frac{7}{16}$ for doing errands, and $\$\frac{11}{16}$ for work: by mistake he paid him $\$1\frac{3}{16}$; how much ought William to return to him?

10. Everett asked his sister if she could tell him how much $4\frac{4}{5}$, plus $3\frac{3}{5}$, plus 2, plus $7\frac{4}{5}$, would equal; she answered 19; how much did her answer differ from the true one?

11. How many are $3\frac{7}{10}+6\frac{4}{10}+8\frac{9}{10}$?

12. How many are $3\frac{2}{11}+\frac{8}{11}+4\frac{4}{11}$?

13. How many are $6\frac{5}{12}+3\frac{11}{12}+2\frac{1}{12}$?

14. How many are $8\frac{6}{7}-3$? $8\frac{6}{7}-3\frac{2}{7}$?

15. How many are $9\frac{10}{11}-\frac{7}{11}$? $9\frac{10}{11}-4\frac{7}{11}$?

16. How many are $1-\frac{1}{2}$? $1-\frac{1}{6}$? $1-\frac{5}{6}$?

17. How many are $2-\frac{1}{2}$? $5-\frac{1}{3}$? $5-\frac{3}{7}$?

18. How many are $5 - 1 - \frac{2}{7}$? $5 - 1\frac{2}{7}$?

19. How many are $17 - 8 - \frac{1}{9}$? $17 - 8\frac{1}{9}$?

20. How many are $13 - 8\frac{2}{9}$? $14 - 13\frac{3}{16}$?

21. How many are $1\frac{1}{3} - \frac{2}{3}$?

NOTE. $1\frac{1}{3} = \frac{4}{3}$; $\frac{4}{3} - \frac{2}{3} = \frac{2}{3}$.

22. How many are $1\frac{4}{9} - \frac{8}{9}$?

23. How many are $1\frac{3}{5} - \frac{4}{5}$?

24. How many are $7\frac{3}{5} - \frac{4}{5}$?

25. How many are $1\frac{5}{7}$? $- \frac{6}{7}$? $8\frac{5}{7} - \frac{6}{7}$?

26. How many are $7\frac{8}{11} - \frac{10}{11}$? $2\frac{4}{13} - \frac{12}{13}$?

27. How many are $6\frac{1}{9} - 4 - \frac{7}{9}$? $6\frac{1}{9} - 4\frac{7}{9}$?

28. How many are $9\frac{3}{8} - 2\frac{5}{8}$? $15\frac{3}{15} - 6\frac{4}{15}$?

29. How many are $5\frac{6}{7} + 2\frac{3}{7} - 5\frac{4}{7}$? $8 - 3\frac{2}{9} + 6\frac{5}{9}$?

59. 1. Charles earned $\$\frac{1}{2}$ on Monday, and $\$\frac{1}{4}$ on Tuesday; how many 4ths of a dollar did he earn on both days?

Solution. He earned $\$\frac{1}{2}$ plus $\$\frac{1}{4}$: $\frac{1}{2}$ equals $\frac{2}{4}$: $\frac{2}{4}$ plus $\frac{1}{4}$ equals $\frac{3}{4}$; therefore, he earned $\$\frac{3}{4}$.

2. Irving has worked $\frac{3}{4}$ of a day for Mr. Joy, and $\frac{1}{2}$ of a day for Mr. Clarke; how many 4ths of a day has he worked for both? how many days?

3. Mary owes $\$\frac{1}{4}$ for paper; if she should give $\$\frac{1}{2}$ in payment, what part of a dollar should be returned to her?

4. Henry gave $\$\frac{3}{8}$ for a fish-line and $\$\frac{1}{2}$ for a fish-pole; how many more 8ths of a dollar did he give for the pole than for the line?

5. Mr. Sweet spends $\frac{1}{2}$ of his time in working, $\frac{1}{8}$ in studying, and $\frac{1}{4}$ in sleeping; what part of his time does he spend in all these ways? What part of his time remains for other purposes?

6. Mrs. Thayer owned $\frac{5}{8}$ of a mill, but she sold $\frac{1}{2}$ of the mill; what part of the mill did she still own?

7. In making a box, a carpenter used $\frac{2}{3}$ of a pound of board nails, $\frac{1}{2}$ of a pound of shingle nails, and $\frac{1}{6}$ of a pound of lath nails; how many pounds did he use?

8. Mr. Williams bought two remnants of silk, one containing $\frac{5}{9}$ of a yard and the other $\frac{1}{3}$ of a yard; what did both remnants contain?

9. A can mow $\frac{2}{5}$ of a field in 1 day, and B can mow $\frac{1}{2}$ of it in 1 day; what part of the field can both mow in 1 day?

Suggestion. Change $\frac{2}{5}$ and $\frac{1}{2}$ to tenths.

10. Mr. Wise bought cloth at $\$\frac{1}{4}$ a yard, and sold it at $\$\frac{1}{3}$ a yard; what did he gain on 1 yard?

11. An estate was divided in the following manner: the widow received $\frac{1}{3}$, the son and two daughters received $\frac{1}{5}$ each, and the rest was required to pay the debts; what part was required to pay the debts?

12. In an orchard $\frac{1}{2}$ of the trees are apple-trees, $\frac{2}{7}$ are pear-trees, and the rest are plum-trees; what part of the orchard is plum-trees?

13. If $\frac{2}{5}$ of a pole is in the water, $\frac{1}{6}$ in the mud, and the rest in the air, what part is in the air?

14. Robert received \$2 for 1 week's work; if he paid $\$\frac{3}{4}$ for a pair of shoes and $\$\frac{2}{5}$ for a hat, how much had he left?

15. A man sold an acre of land for $\frac{9}{10}$ of what it cost, and thereby lost \$12; what part of the cost did he lose? What did the land cost?

16. The difference between $\frac{1}{2}$ of my age and $\frac{2}{3}$ of my age is 4 years; what is my age?

17. After $\frac{4}{5}$ and $\frac{1}{6}$ of a flock of sheep are sold, 3 sheep remain; what number of sheep were in the flock at first?

18. If it requires $2\frac{2}{5}$ hours to plough a garden, and $11\frac{1}{4}$ hours to plant it, how much more time is required to plant the garden than to plough it?

19. Edwin had $\$\frac{7}{8}$ and William had $\$\frac{1}{2}$: after Edwin

had spent $\$\frac{3}{4}$, and William had spent $\$\frac{5}{16}$, how much less had Edwin than William?

20. What number must be added to $\frac{3}{4}$ plus $\frac{5}{6}$ to make 2 units?

21. What number must be taken out of $7\frac{1}{2}$ to leave $3\frac{2}{3}$?

22. How many are $8\frac{3}{14}+2\frac{6}{7}$? $14\frac{1}{7}-6\frac{1}{2}$?

23. How many are $4\frac{5}{6}+2\frac{3}{8}$? $15\frac{3}{4}-6\frac{4}{5}$?

24. How many are $20\frac{1}{4}+17\frac{9}{10}$? $13\frac{1}{3}-12\frac{1}{2}$?

Section XXVII.

MULTIPLICATION OF FRACTIONS BY WHOLE NUMBERS.

60. 1. If James earns 1 third of a dollar in 1 day, how much will he earn in 2 days?

2. 2 times 1 third (or two 1's of thirds), are how many thirds?

3. If it takes Dana 1 third of an hour to walk 1 mile, in what time can he walk 3 miles?

4. 3 times 1 third (or three 1's of thirds), are how many 1's?

5. If Warren gives 1 half of a melon to each of 5 boys, how many halves does he give to all? how many melons?

Solution. If he gives 1 half of a melon to each boy, to 5 boys he will give 5 times 1 half of a melon, or 5 halves, which equals 2 melons and 1 half; therefore, etc.

6. Oscar gave $\frac{1}{3}$ of an apple to each of 7 boys; how many apples did he give away?

7. 7 times 1 third (or seven 1's of thirds), are how many 1's?

8. Alonzo gave $\$\frac{1}{3}$ for his hat; what would 8 hats cost at the same rate?

9. Annie knits $\frac{1}{4}$ of a mitten in 1 day; how many mittens can she knit in 8 days?

10. How many are 7 times $\frac{1}{4}$? 8 times $\frac{1}{5}$?

11. How many are 10 times $\frac{1}{6}$? 11 times $\frac{1}{9}$?

12. How many are 12 times $\frac{1}{8}$? 12 times $\frac{1}{7}$?

13. Horace can weed $\frac{2}{3}$ of a row of plants in 1 hour; how many rows can he weed in 2 hours? in 7 hours?

14. How many are 2 times $\frac{2}{3}$? 7 times $\frac{2}{3}$?

15. How many are 2 times $\frac{3}{4}$? 8 times $\frac{3}{4}$?

16. How many yards are there in 3 aprons, if each apron contains $\frac{3}{5}$ of a yard?

17. If a man receives $\$\frac{3}{4}$ for shoeing 1 horse, what will he receive for shoeing 6 horses?

18. If $\frac{5}{8}$ of a pound of cheese will last a family 1 day, how many pounds will last them 7 days? 12 days?

19. If 3 men can do a piece of work in $\frac{2}{3}$ of a day, in what time can 1 man do the same work?

20. How many are 7 times $\frac{4}{5}$? 9 times $\frac{4}{6}$?

21. How many are 8 times $\frac{2}{7}$? 11 times $\frac{5}{8}$?

22. How many are 6 times $\frac{2}{9}$? 3 times $\frac{1}{12}$?

23. How many are 10 times $\frac{7}{8}$? 4 times $\frac{3}{11}$?

24. A grocer gains $\$\frac{2}{5}$ upon a pound of nutmegs, which is 1 fourth of the cost; what is the cost?

25. $\frac{2}{5}$ is 1 fourth of what number?

26. $\frac{3}{8}$ is 1 third of what number?

61. 1. If it takes $1\frac{1}{2}$ yards of cloth to make 1 cape, how many yards will it take to make 3 capes?

Solution. If it takes $1\frac{1}{2}$ yards to make 1 cape, to make 3 capes it will take 3 times $1\frac{1}{2}$ yards: 3 times 1 are 3, and 3 times $\frac{1}{2}$ are $\frac{3}{2}$, or $1\frac{1}{2}$, which, with 3, equals $4\frac{1}{2}$; therefore, etc.

2. If 1 pound of crackers costs $12\frac{1}{2}$ cents, what will 4 pounds cost?

3. How many are 4 times $12\frac{1}{2}$? 3 times $2\frac{2}{3}$?

4. Emma has $\$2\frac{3}{5}$, and when she gets twice as much, she can buy a dress; how much will the dress cost?

5. How many are 2 times $2\frac{3}{5}$? 4 times $4\frac{5}{6}$?

6. How many are 7 times $1\frac{2}{7}$? 8 times $3\frac{4}{9}$?

7. If 1 box of lemons costs $\$4\frac{3}{8}$, what will 3 boxes cost? 4 boxes? 7 boxes?

8. If a horse can trot a mile in $2\frac{3}{4}$ minutes, in how many minutes can he trot 3 miles? 5 miles?

9. If 2 men can shingle a roof in $2\frac{2}{3}$ days, in how many days can 1 man shingle it?

10. If a lot of cloth can be woven in $4\frac{1}{2}$ days upon 2 looms, in how many days can it be woven upon 1 loom?

11. If a case of boots can be made by 4 men in $2\frac{1}{12}$ weeks, in how many weeks can 1 man do the same work?

12. How many are 4 times $2\frac{3}{4}$? 6 times $3\frac{2}{5}$?

13. How many are 8 times $5\frac{1}{7}$? 10 times $3\frac{2}{9}$?

14. How many are 9 times $5\frac{4}{5}$? 11 times $7\frac{3}{8}$?

15. 4 men bought a boat together, each person paying as his share $\$15\frac{5}{8}$; what was the price of the boat?

16. Mr. B. paid a tax of $\$13\frac{1}{2}$ last year, which was $\frac{1}{4}$ of what Mr. E. paid; what did Mr. E. pay?

17. $8\frac{2}{3}$ is $\frac{1}{5}$ of what number?

18. $20\frac{3}{4}$ is $\frac{1}{3}$ of what number?

Section XXVIII.

MULTIPLICATION OF WHOLE NUMBERS BY FRACTIONS.

62. 1. If 1 loaf of bread is worth 8 cents, what is $\frac{1}{2}$ of a loaf worth?

2. If 1 yard of ribbon is worth 12 cents, what is $\frac{1}{3}$ of a yard worth? what are $\frac{2}{3}$ of a yard worth?

Solution. If 1 yard is worth 12 cents, $\frac{1}{3}$ of a yard is worth $\frac{1}{3}$ of 12 cents, which is 4 cents, and $\frac{2}{3}$ of a yard are worth 2 times 4 cents, or 8 cents; therefore, etc.

3. In a score there are 20 things; how many things are there in $\frac{1}{4}$ of a score? in $\frac{3}{4}$ of a score?

4. When coal is worth $10 a ton, what is $\frac{1}{5}$ of a ton worth? what are $\frac{3}{5}$ of a ton worth?

5. If a field of 18 acres can be reaped in a week, how many acres can be reaped in $\frac{1}{6}$ of a week? in $\frac{3}{6}$? in $\frac{5}{6}$?

6. If a man can travel 42 miles in a day, how many miles can he travel in $\frac{3}{7}$ of a day? in $\frac{5}{6}$ of a day?

7. In $1 there are 100 cents; how many cents are there in $\$\frac{1}{10}$? in $\$\frac{2}{10}$? in $\$\frac{7}{10}$? in $\$\frac{9}{10}$? in $\$\frac{1}{5}$? in $\$\frac{3}{4}$?

8. There are 144 things in a gross; how many things are there in $\frac{5}{12}$ of a gross? in $\frac{7}{12}$?

9. How many buttons are there in $\frac{11}{12}$ of a gross of buttons?

63. 1. If a clerk can copy 5 pages in 1 hour, how many pages can he copy in $\frac{1}{3}$ of an hour?

Solution. If he can copy 5 pages in 1 hour, in $\frac{1}{3}$ of an hour he can copy $\frac{1}{3}$ of 5 pages: 5 pages equals 3 pages plus 2 pages; $\frac{1}{3}$ of 3 is 1, and $\frac{1}{3}$ of 2 is $\frac{2}{3}$, which, with 1, equals $1\frac{2}{3}$; therefore, etc.

2. If a wheel turns 10 times in going 1 rod, how many times will it turn in going $\frac{1}{4}$ of a rod?

3. If a man can do a piece of work in 35 days, in how many days can he do $\frac{1}{4}$ of it?

4. If a barrel of flour costs $14, what will $\frac{1}{5}$ of a barrel cost? what will $\frac{3}{5}$ of a barrel cost?

5. What is $\frac{1}{5}$ of 14? $\frac{3}{5}$ of 14? $\frac{3}{5}$ of 37?

6. If 1 acre of land costs $44, what must be paid for $\frac{1}{6}$ of an acre? for $\frac{5}{6}$ of an acre?

7. What is $\frac{1}{6}$ of 44? $\frac{5}{6}$ of 44? $\frac{5}{6}$ of 39?

8. If a peck of peas costs 50 cents, what will $\frac{1}{8}$ of a peck cost? $\frac{3}{8}$ of a peck? $\frac{5}{8}$ of a peck?

9. What is $\frac{1}{8}$ of 50? $\frac{5}{8}$ of 50? $\frac{4}{9}$ of 70?

10. What is $\frac{1}{7}$ of 38? $\frac{2}{7}$ of 38? $\frac{4}{10}$ of 63?

11. What is $\frac{2}{3}$ of 25? $\frac{5}{6}$ of 25? $\frac{5}{12}$ of 86?

12. What is $\frac{3}{11}$ of 48? $\frac{9}{11}$ of 48? $\frac{7}{12}$ of 52?

13. What cost $3\frac{1}{2}$ quarts of cherries at 8 cents a quart?

14. What cost $5\frac{2}{3}$ quarts of cherries at 9 cents a quart?

15. What cost $8\frac{3}{4}$ yards of cloth at $8 a yard?

16. What cost $8\frac{1}{21}$ yards of silk at $11 a yard?

17. What cost $10\frac{3}{5}$ bushels of cranberries at $3 a bushel?

18. If a horse can travel 1 mile in 9 minutes, in how many minutes can he travel $3\frac{4}{7}$ miles?

19. John can earn $7 in 1 week; how many dollars can he earn in $6\frac{2}{3}$ weeks?

20. Thomas can earn 12 cents in 1 hour; how much can he earn in $6\frac{3}{4}$ hours?

21. How many are $7\frac{4}{10}$ times 6? $2\frac{5}{6}$ times 11?

22. How many are $11\frac{4}{9}$ times 4? $7\frac{6}{11}$ times 10?

Section XXIX.

MULTIPLICATION OF FRACTIONS BY FRACTIONS.

64.

$\frac{1}{2}$ $\frac{1}{4}$ 1. If $\frac{1}{2}$ of an inch is divided into 2 equal parts, what part of an inch is one of the parts?

2. If $\frac{1}{2}$ of a melon is divided into 2 equal parts, what part of the melon is one of the parts?

Answer and Solution. $\frac{1}{4}$ of a melon; because, if each half of a melon were divided into 2 equal parts, the whole of the melon would be divided into 2 times 2, or 4 equal parts; therefore, 1 of the parts is $\frac{1}{4}$ of the melon.

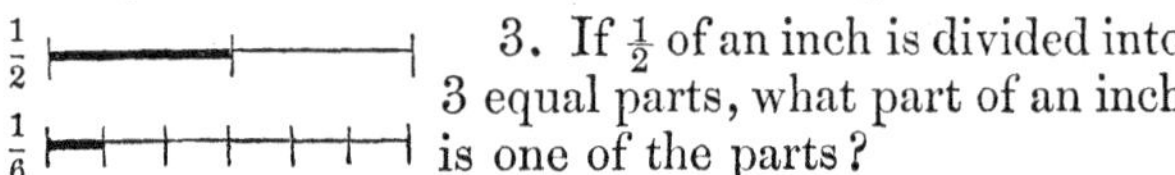

3. If $\frac{1}{2}$ of an inch is divided into 3 equal parts, what part of an inch is one of the parts?

4. What is $\frac{1}{3}$ of $\frac{1}{2}$ of a pear?

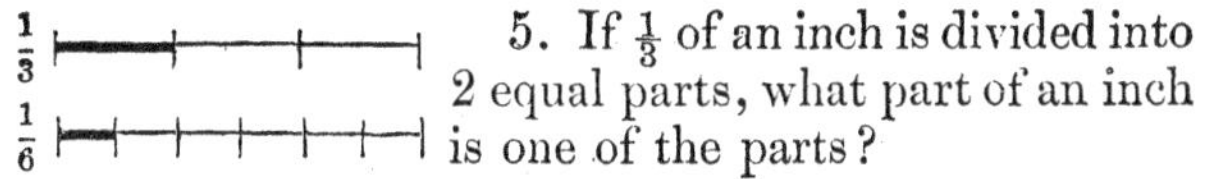

5. If $\frac{1}{3}$ of an inch is divided into 2 equal parts, what part of an inch is one of the parts?

6. What is $\frac{1}{2}$ of $\frac{1}{3}$ of a lemon? $\frac{1}{3}$ of $\frac{1}{2}$?

7. What is $\frac{1}{2}$ of $\frac{1}{4}$? $\frac{1}{4}$ of $\frac{1}{2}$? $\frac{1}{2}$ of $\frac{1}{5}$?

8. What is $\frac{1}{5}$ of $\frac{1}{2}$? $\frac{1}{2}$ of $\frac{1}{6}$? $\frac{1}{6}$ of $\frac{1}{2}$?

9. John had $\frac{1}{3}$ of a quart of nuts, and gave away $\frac{1}{3}$ of what he had; what part of a quart did he give away?

10. What is $\frac{1}{3}$ of $\frac{1}{3}$ of a quart? Why?

11. What is $\frac{1}{3}$ of $\frac{1}{4}$? $\frac{1}{3}$ of $\frac{1}{7}$? $\frac{1}{3}$ of $\frac{1}{9}$?

12. Mr. Jones had $\frac{1}{3}$ of a bushel of seed potatoes, and gave $\frac{1}{4}$ of what he had to Mr. Gray; what part of a bushel did he give to Mr. Gray?

13. What is $\frac{1}{4}$ of $\frac{1}{3}$? $\frac{1}{4}$ of $\frac{1}{6}$? $\frac{1}{4}$ of $\frac{1}{7}$?

14. What is $\frac{1}{7}$ of $\frac{1}{2}$? $\frac{1}{8}$ of $\frac{1}{6}$? $\frac{1}{11}$ of $\frac{1}{4}$?

15. Mrs. Fogg had $\frac{3}{5}$ of a barrel of soap, and used $\frac{1}{2}$ of what she had; what part of the barrel did she use?

Solution. She used $\frac{1}{2}$ of $\frac{3}{5}$ of a barrel; $\frac{1}{2}$ of $\frac{1}{5}$ is $\frac{1}{10}$, and $\frac{1}{2}$ of $\frac{3}{5}$ is 3 times $\frac{1}{10}$, or $\frac{3}{10}$; therefore, etc.

16. Mr. Rice owned $\frac{5}{6}$ of an acre of land, and sold $\frac{1}{3}$ of what he owned; what part of an acre did he sell?

17. A woman had $\frac{7}{8}$ of a yard of silk, and used $\frac{1}{3}$ of it; what part of a yard did she use?

18. Charles had $\frac{3}{4}$ of a cake, and gave $\frac{1}{5}$ of it to Henry; what part of the cake did he give to Henry?

19. If a bushel of tomatoes costs $\$\frac{7}{8}$, what will $\frac{1}{6}$ of a bushel cost? $\frac{1}{8}$ of a bushel?

20. What is $\frac{1}{3}$ of $\frac{2}{3}$? $\frac{1}{3}$ of $\frac{2}{5}$? $\frac{1}{4}$ of $\frac{3}{7}$?

21. What is $\frac{1}{4}$ of $\frac{5}{8}$? $\frac{1}{8}$ of $\frac{3}{5}$? $\frac{1}{7}$ of $\frac{2}{3}$?

22. What is $\frac{1}{10}$ of $\frac{4}{9}$? $\frac{1}{5}$ of $\frac{2}{11}$? $\frac{1}{6}$ of $\frac{3}{4}$?

23. From what you have learned in this section, how can you find $\frac{1}{2}$ of any fractional number? *Ans.* **By multiplying the denominator by 2.**

24. In what other way can you find $\frac{1}{2}$ of a fractional number? *Ans.* By dividing the numerator by 2.

25. Why is the result the same by both operations?

26. In what two ways can you find $\frac{1}{3}$ of a fractional number? $\frac{1}{5}$? $\frac{1}{7}$? $\frac{1}{8}$?

NOTE. In obtaining part of a fractional number, it is better to divide the numerator, when it can be divided without leaving a remainder, as the fractional number is then expressed in smaller terms.

27. Jane had $\frac{3}{4}$ of a yard of ribbon, and made a needle-book with $\frac{1}{3}$ of what she had; what part of a yard did she use?

28. What will $\frac{1}{4}$ of a pound of butter cost at $\$\frac{4}{10}$ a pound? What will $\frac{3}{4}$ of a pound cost?

Solution. If 1 pound of butter costs $\$\frac{4}{10}$, $\frac{1}{4}$ of a pound will cost $\frac{1}{4}$ of $\$\frac{4}{10}$, or $\$\frac{1}{10}$, and $\frac{3}{4}$ of a pound will cost 3 times $\$\frac{1}{10}$, or $\$\frac{3}{10}$; therefore, etc.

29. What cost $\frac{4}{5}$ of a pound of figs at $\$\frac{5}{16}$ a pound?

30. A paper-hanger had $\frac{6}{7}$ of a barrel of flour, and made $\frac{5}{6}$ of it into paste; what part of a barrel did he make into paste?

31. What cost $\frac{2}{3}$ of a yard of linen at $\$\frac{3}{4}$ a yard?

32. What cost $\frac{3}{4}$ of a yard of calico at $\$\frac{2}{3}$ a yard?

33. What cost $\frac{5}{6}$ of a dozen hooks at $\$\frac{4}{5}$ a dozen?

34. If a man can earn $\$\frac{3}{8}$ in 1 hour, how much can he earn in $\frac{3}{5}$ of an hour?

35. A tailor, having $\frac{2}{3}$ of a bale of cloth, sold $\frac{3}{7}$ of what he had; what part of the bale did he sell?

36. If a woman can do $\frac{9}{10}$ of a piece of work in a day, how much can she do in $\frac{4}{5}$ of a day?

37. John had $\frac{8}{9}$ of a bushel of apples, and sold $\frac{3}{4}$ of what he had; what part of a bushel did he sell?

38. Carrie put $\frac{7}{8}$ of a pound of flour into some cake; how much flour will she require to make $\frac{4}{9}$ as much cake?

39. What is $\frac{3}{4}$ of $\frac{2}{7}$? $\frac{2}{7}$ of $\frac{2}{5}$? $\frac{4}{3}$ of $\frac{3}{7}$?

40. What is $\frac{3}{8}$ of $\frac{3}{10}$? $\frac{1}{8}$ of $\frac{2}{11}$? $\frac{4}{9}$ of $\frac{5}{6}$?

Section XXX.

MULTIPLICATION OF MIXED NUMBERS BY FRACTIONS, &c

65. 1. If Homer's spelling-book cost him $12\frac{1}{2}$ cents, and his pencil cost him $\frac{2}{5}$ as much, what did his pencil cost?

Solution. If his spelling-book cost him $12\frac{1}{2}$ cents, and his pencil cost him $\frac{2}{5}$ as much, his pencil must have cost him $\frac{2}{5}$ of $12\frac{1}{2}$ cents: $12\frac{1}{2}$ equals 10 plus $2\frac{1}{2}$; $\frac{1}{5}$ of 10 equals 2, and $\frac{1}{5}$ of $2\frac{1}{2}$, or $\frac{5}{2}$, equals $\frac{1}{2}$, which, with 2, equals $2\frac{1}{2}$. If $2\frac{1}{2}$ is $\frac{1}{5}$ of some number, $\frac{2}{5}$ of the number must be 2 times $2\frac{1}{2}$, or 5; therefore, etc.

2. Mary's bonnet cost her $\$6\frac{2}{3}$, and Sarah's cost $\frac{3}{5}$ as much; what did Sarah's cost?

3. What is $\frac{3}{5}$ of $6\frac{2}{3}$? $\frac{5}{6}$ of $8\frac{4}{7}$? $\frac{2}{7}$ of $15\frac{1}{6}$?

4. What is $\frac{3}{7}$ of $25\frac{3}{8}$? $\frac{9}{10}$ of $7\frac{4}{8}$? $\frac{8}{9}$ of $4\frac{1}{2}$?

5. A boy, who had $27\frac{1}{5}$ miles to travel, rode $\frac{3}{8}$ of the distance, and walked the rest; how many miles did he ride? How many miles did he walk?

6. What is $\frac{3}{8}$ of $27\frac{1}{5}$? $\frac{5}{8}$ of $28\frac{4}{5}$? $\frac{5}{7}$ of $31\frac{1}{9}$?

7. What is $\frac{7}{10}$ of $54\frac{2}{7}$? $\frac{4}{9}$ of $20\frac{4}{7}$? $\frac{3}{11}$ of $13\frac{4}{9}$?

66. 1. When 1 pound of cotton is required to make $3\frac{1}{2}$ yards of cloth, how many yards can be made from $\frac{1}{3}$ of a pound of cotton? from $\frac{2}{3}$ of a pound?

Suggestion. $3\frac{1}{2}=3+\frac{1}{2}$: $\frac{1}{3}$ of 3 is 1; $\frac{1}{3}$ of $\frac{1}{2}$ is $\frac{1}{6}$, etc.

2. If a box of raisins costs $\$4\frac{3}{4}$, what will $\frac{1}{4}$ of a box cost? $\frac{3}{4}$ of a box?

3. What is $\frac{1}{4}$ of $4\frac{3}{4}$? $\frac{3}{4}$ of $4\frac{3}{4}$? $\frac{3}{4}$ of $16\frac{2}{3}$?

4. If a horse travels $10\frac{1}{4}$ miles in 1 hour, how many miles does he travel in $\frac{1}{5}$ of an hour? in $\frac{2}{5}$ of an hour?

5. What is $\frac{1}{5}$ of $10\frac{1}{4}$? $\frac{2}{5}$ of $10\frac{1}{4}$? $\frac{3}{5}$ of $40\frac{3}{4}$?

6. A mother is $32\frac{1}{6}$ years old, and her eldest son is $\frac{3}{8}$ as old as she is; how old is the son?

7. If a barrel of apples is sold for $\$1\frac{1}{2}$, what is the price of $\frac{1}{2}$ of a barrel?

NOTE. Change $\$1\frac{1}{2}$ to halves.

8. A man had $\$2\frac{1}{3}$ in his purse, and spent $\frac{1}{4}$ of it; how much did he spend?

9. What is $\frac{1}{4}$ of $2\frac{1}{3}$? $\frac{1}{4}$ of $3\frac{2}{5}$? $\frac{1}{5}$ of $2\frac{4}{5}$?

10. A family put up $4\frac{1}{3}$ barrels of potatoes for the winter, and ate $\frac{1}{7}$ of them in a month; what quantity of potatoes did they eat in a month?

11. If $6\frac{1}{7}$ yards of ribbon can be bought for \$1, how much can be bought for $\$\frac{1}{8}$?

12. If the wages of a teacher are $\$9\frac{1}{2}$ a week, what are her wages for $\frac{1}{2}$ of a week?

13. If board is $\$3\frac{1}{2}$ a week, what is a person's board for $\frac{1}{2}$ of a week?

14. What is $\frac{1}{2}$ of $3\frac{1}{2}$? of $5\frac{1}{2}$? of $7\frac{1}{2}$?

15. What is $\frac{1}{2}$ of $11\frac{1}{2}$? of $13\frac{1}{2}$? of $15\frac{1}{2}$?

16. What is $\frac{1}{3}$ of $4\frac{1}{3}$ bushels of wheat?

17. If a man walks $4\frac{1}{4}$ miles in 1 hour, how far can he walk in $\frac{1}{3}$ of an hour? in $\frac{2}{3}$ of an hour?

18. At $\$8\frac{1}{4}$ a ream, what cost $\frac{1}{3}$ of a ream of paper?

19. What is $\frac{2}{3}$ of $\frac{1}{4}$? $\frac{3}{4}$ of $9\frac{1}{2}$?

20. A train of cars went at the rate of $17\frac{1}{3}$ miles in an hour; how far did it go in $\frac{3}{4}$ of an hour?

21. What is $\frac{3}{4}$ of $17\frac{1}{3}$? $\frac{5}{2}$ of $20\frac{1}{6}$?

22. Louis is $18\frac{1}{3}$ years old, and Sarah is $\frac{3}{5}$ as old as Louis; how old is Sarah?

23. A man bought a cow for $\$37\frac{1}{2}$, and sold her again for $\frac{5}{6}$ of her cost; for how much did he sell her?

24. If a barrel of peas sells for $\$10\frac{1}{2}$, for what will $\frac{1}{2}$ of a barrel sell? $1\frac{1}{2}$ barrels?

NOTE. Change $1\frac{1}{2}$ to halves.

25. If a man can build $12\frac{1}{2}$ rods of fencing in 1 day, how many rods can he build in $\frac{1}{3}$ of a day? in $1\frac{1}{3}$ days?

NOTE. Change $1\frac{1}{3}$ to thirds.

26. What is $1\frac{1}{3}$ times $12\frac{1}{2}$? $1\frac{2}{3}$ times $9\frac{1}{2}$?

27. If $1 will pay for $13\frac{1}{2}$ pounds of sugar, how much sugar can be bought for $\$\frac{1}{2}$? for $\$2\frac{1}{2}$?

28. What is $2\frac{1}{2}$ times $13\frac{1}{2}$? $3\frac{1}{2}$ times $20\frac{1}{2}$?

29. If 1 ton of hay costs $\$17\frac{3}{4}$, what will $\frac{2}{5}$ of a ton cost at the same rate?

30. A woman is $22\frac{1}{3}$ years old: she has a sister who is $\frac{1}{3}$ as old as she is, and a brother who is 5 times as old as her sister; how old is her sister? How old is her brother?

31. What is $\frac{5}{3}$ of $22\frac{1}{3}$? $\frac{7}{3}$ of $14\frac{1}{3}$?

32. If a horse travels $7\frac{2}{5}$ miles in 1 hour, and a locomotive goes $3\frac{1}{2}$ times as fast as the horse, how many miles does the locomotive go in 1 hour?

67. 1. I bought 3 table-spoons at $\$3\frac{4}{5}$ each, and 2 dozen teaspoons at $\$8\frac{3}{5}$ a dozen; what must I pay for both lots of spoons?

2. If the upper leather for a pair of shoes costs $\$\frac{2}{5}$, the sole leather $\$\frac{1}{5}$, and the making $\$\frac{2}{5}$, what will be the whole cost of 1 pair? of 2 pairs? of 4 pairs?

3. What cost 6 umbrellas at $\$2\frac{2}{3}$ apiece? How much silk at $2 per yard will be required to pay for the umbrellas?

4. If 5 men can do a piece of work in $3\frac{3}{4}$ days, in how many days can 1 man do it? How many more days would be required for 1 man than for 5 to do it?

5. Mr. Ward purchased 6 chairs at $\$\frac{3}{4}$ apiece, and 2 tables at $\$7\frac{3}{4}$ apiece; how much should he pay for the chairs? for the tables? for both chairs and tables? Having paid but $10, he gave his note for the rest; for how much was the note given?

Section XXXI.

DIVISION BY WHOLE NUMBERS.

68. 1. If 1 yard of ribbon costs 4 cents, what part of a yard can be bought for 1 cent? for 2 cents? for 3 cents?

2. What part of 4 is 1? is 2? is 3?

3. If 1 yard of ribbon costs 4 cents, how many yards can be bought for 11 cents? *

Solution I. If 1 yard costs 4 cents, as many yards can be bought for 11 cents as there are 4's in 11: in 11 there are two 4's and $\frac{3}{4}$ of 4; therefore, $2\frac{3}{4}$ yards can be bought for 11 cents.

Solution II. If 1 yard costs 4 cents, as many yards can be bought for 11 cents as there are times 4 in 11, which is $2\frac{3}{4}$ times; therefore, etc.

4. At 6 cents a pound, how many pounds of tomatoes can be bought for 13 cents?

5. 13 are how many 6's? *Ans.* $2\frac{1}{6}$.

6. 14 are how many 3's?

7. 16 are how many 3's? 5's? 6's? 7's?

8. 19 are how many 5s'? 6's? 8's? 9's?

9. 22 are how many 3's? 5's? 8's? 4's?

10. If Mary can make 8 button-holes in 1 hour, in how many hours can she make 22 button-holes? 29 button-holes?

11. How many pounds of rice at 7 cents a pound can be bought for 38 cents?

12. How many pounds of sugar at 9 cents a pound can be bought for 40 cents?

13. How many cords of wood at $6 a cord can be bought for $65?

* *Analytical Solution.* Since 4 cents will buy 1 yard, 1 cent will buy $\frac{1}{4}$ of a yard, and 11 cents will buy $\frac{11}{4}$, which equals $2\frac{3}{4}$ yards; therefore, etc.

69. 1. If 2 books cost $\$\frac{4}{5}$, what will 1 book cost?

Solution. If 2 books cost $\$\frac{4}{5}$, 1 book will cost 1 half of $\$\frac{4}{5}$, or $\$\frac{2}{5}$; therefore, etc.

2. If 3 yards of cloth cost $\$\frac{9}{10}$, what will 1 yard cost?

3. If 1 man can do a piece of work in $\frac{6}{7}$ of a day, how long will it take 2 men to do it?

4. $\frac{4}{5}$ divided by 2 equals what?

5. $\frac{9}{10}$ divided by 3 equals what?

6. $\frac{8}{9}$ divided by 4 equals what?

7. If 4 loaves of bread can be bought for $\$\frac{1}{2}$, what will 1 loaf cost?

Solution. If 4 loaves can be bought for $\$\frac{1}{2}$, 1 loaf will cost 1 fourth of $\$\frac{1}{2}$, or $\$\frac{1}{8}$; therefore, etc.

8. If 5 men can plant $\frac{1}{3}$ of a piece of land in a day, what part of the piece can 1 man plant?

9. Divide $\frac{1}{2}$ by 4; $\frac{1}{3}$ by 5; $\frac{3}{4}$ by 3.

10. In what two ways can you divide a fractional number by a whole number? [See Props. II. and III., p. 72.]

NOTE. In dividing a fractional number by a whole number, it is better to divide the numerator, whenever it can be divided without a remainder, as the answer is then expressed in smaller terms.

11. Divide $\frac{1}{7}$ by 3; $\frac{4}{7}$ by 3; $\frac{3}{7}$ by 3.

12. Divide $\frac{1}{8}$ by 2; $\frac{4}{9}$ by 2; $\frac{3}{6}$ by 2.

13. Divide $5\frac{1}{2}$ by 5; $6\frac{1}{4}$ by 5; $8\frac{3}{4}$ by 5.

14. Divide $9\frac{1}{3}$ by 6; $10\frac{1}{2}$ by 7; $15\frac{1}{8}$ by 8.

15. Divide $8\frac{7}{10}$ by 6; $4\frac{1}{9}$ by 10; $11\frac{4}{5}$ by 11.

16. Divide $23\frac{2}{5}$ by 9; $17\frac{1}{2}$ by 11; $42\frac{8}{9}$ by 7.

17. If 4 months is $\frac{1}{3}$ of a year, what part of a year is 1 month?

18. If in 3 rods there are $16\frac{1}{2}$ yards, how many yards are there in 1 rod?

19. If the railroad fare for 5 persons is $\$31\frac{1}{4}$, what is the fare for 1 person?

20. If $10\frac{2}{5}$ rods of wall were laid for $7, how much could be laid for $1?

Section XXXII.

Division of Fractions by Fractions of the same Denominators.

70. 1. James had $\frac{6}{9}$ of a melon; to how many boys can he give $\frac{2}{9}$ of a melon apiece?

Solution. If he gives $\frac{2}{9}$ of a melon to each boy, he can give $\frac{6}{9}$ to as many boys as there are 2's of 9ths in $\frac{6}{9}$, or 2's in 6; there are three 2's in 6; therefore, etc.

2. $\frac{6}{9}$ are how many times $\frac{2}{9}$?

Solution. $\frac{6}{9}$ are as many times $\frac{2}{9}$, as there are 2's of ninths in $\frac{6}{9}$, or 2's in 6, etc.

3. $\frac{9}{10}$ are how many times $\frac{3}{10}$? $\frac{2}{10}$? $\frac{7}{10}$?

4. $\frac{10}{11}$ are how many times $\frac{2}{11}$? $\frac{5}{11}$? $\frac{6}{11}$?

5. A shoemaker has contracted a debt of \$$2\frac{4}{7}$, which he wishes to discharge by making boots at \$$\frac{6}{7}$ per pair; how many pairs of boots must he make?

Note. Change $2\frac{4}{7}$ to sevenths.

6. How many fans at \$$\frac{2}{5}$ apiece can be bought for \$$3\frac{1}{5}$?

7. How many garments can be made from $6\frac{3}{4}$ yards of cassimere, if it takes $2\frac{1}{4}$ yards for each garment?

8. $3\frac{3}{4}$ are how many times $2\frac{1}{4}$? $\frac{3}{4}$? $1\frac{3}{4}$?

9. $7\frac{1}{3}$ are how many times $3\frac{2}{3}$? $\frac{2}{3}$? $2\frac{1}{3}$?

10. $12\frac{5}{6}$ are how many times $1\frac{5}{6}$? $1\frac{1}{6}$? $1\frac{2}{6}$?

11. If a boy can saw $\frac{4}{7}$ of a cord of wood in 1 day, in how many days can he saw 1 cord?

Note. 1 cord is equal to $\frac{7}{7}$ of a cord. *Ans.* $1\frac{3}{4}$.

12. If a man can do $\frac{2}{11}$ of a certain piece of work in 1 day, in how many days can he do the whole work?

13. 1 is how many times $\frac{3}{5}$? $\frac{8}{9}$? $\frac{11}{12}$?

14. 1 is how many times $\frac{1}{4}$? $\frac{5}{11}$? $\frac{6}{7}$?

15. 1 is how many times $\frac{3}{8}$? $\frac{5}{8}$? $\frac{2}{3}$?

16. If a man can hoe $1\frac{2}{3}$ rows of corn in an hour, in how many hours can he hoe 6 rows?

NOTE. Change 6 to thirds.

17. How many are $6 \div 1\frac{2}{3}$? *Ans.* $3\frac{3}{5}$.
18. How many are $8 \div 1\frac{2}{7}$? $12\frac{1}{5} \div 1\frac{1}{5}$?
19. How many are $8\frac{2}{3} \div 2\frac{1}{3}$? $9\frac{1}{4} \div 1\frac{1}{4}$?
20. How many are $\frac{5}{8} \div 1\frac{3}{8}$ *Ans.* $\frac{5}{11}$.
21. How many are $\frac{7}{12} \div \frac{9}{12}$? $\frac{8}{11} \div 1\frac{3}{11}$?
22. How many are $5\frac{1}{7} \div 1\frac{2}{7}$? $1\frac{9}{15} \div \frac{8}{15}$?
23. How can you divide one fraction by another when the denominators are alike?
24. At $\$\frac{1}{4}$ a pound, how many pounds of tea can be bought for $\$\frac{7}{8}$? *

NOTE. Change $\frac{1}{4}$ to eighths.

25. How many yards of muslin at $\$\frac{1}{5}$ a yard can be purchased for $\$\frac{9}{10}$?
26. How many bushels of apples at $\$\frac{2}{3}$ a bushel can be bought for $\$4\frac{1}{2}$?
27. If 1 pound of nutmegs costs $\$\frac{4}{5}$, what part of a pound can be bought for $\$\frac{1}{5}$? for $\$\frac{2}{5}$? for $\$\frac{3}{5}$?
28. If a bushel of wheat can be bought for $2, how much can be bought for $\$\frac{3}{5}$?

NOTE. Change $2 to fifths.

29. If a barrel of flour costs $6, how much can be bought for $\$2\frac{1}{2}$?
30. If a woman can stitch a vest in 2 hours, how many vests can she stitch in $3\frac{3}{4}$ hours?
31. If a row of corn can be planted in $1\frac{1}{6}$ hours, how many rows can be planted in $5\frac{1}{2}$ hours?

* *Analytical Solution.* Since $\$\frac{1}{4}$ will buy 1 pound, $1 will buy 4 pounds, and $\$\frac{7}{8}$ will buy $\frac{7}{8}$ of 4 pounds; $\frac{7}{8}$ of 4 pounds is $\frac{28}{8}$, which equals $3\frac{1}{2}$ pounds; therefore, etc.

Section XXXIII.

Division of Whole Numbers and Fractions by Fractions.

71. 1. How many lilies at $\frac{1}{2}$ of a cent apiece can be bought for 2 cents?

Solution. As many lilies at $\frac{1}{2}$ of a cent apiece can be bought for 2 cents as there are 1's of halves in 2; 2 changed to halves equals $\frac{4}{2}$; there are as many 1's of halves in $\frac{4}{2}$ as there are 1's in 4; there are four 1's in 4; therefore 4 lilies can be bought.

2. James has 3 cords of wood to saw; if he saws $\frac{1}{3}$ of a cord in each day, in how many days will he saw the whole?

3. How many times $\frac{1}{3}$ are there in 3? in 4? in 5?

Solution. 3 equals $\frac{9}{3}$; there are as many times $\frac{1}{3}$ in 3 as there are 1's of thirds in $\frac{9}{3}$, or 1's in 9; there are nine 1's in 9; therefore there are 9 times $\frac{1}{3}$ in 3.

4. How many times $\frac{1}{4}$ are there in 2? in 3? in 9?

5. George had 2 cords of wood to split for kindlings: if he split $\frac{1}{5}$ of a cord in each week, in how many weeks did he split the whole?

6. How many times $\frac{1}{5}$ are there in 2? in 5? in 7?

7. Dora made 4 yards of linen into collars and cuffs, putting $\frac{1}{7}$ of a yard into each set; how many sets did she make?

8. 4 are how many times $\frac{1}{5}$? $\frac{1}{7}$? $\frac{1}{8}$?

9. 5 are how many times $\frac{1}{6}$? $\frac{1}{9}$? $\frac{1}{10}$?

10. How many are 11 divided by $\frac{1}{6}$? $12 \div \frac{1}{9}$?

11. How many are 9 divided by $\frac{1}{3}$? $15 \div \frac{1}{2}$?

12. How many pecks of tomatoes at $\$\frac{2}{5}$ a peck can be bought for $3?

Solution. As many pecks at $\$\frac{2}{5}$ a peck can be bought for $3 as there are 2's of fifths in 3; 3 changed to fifths equals $\frac{15}{5}$; there are as many 2's of fifths in $\frac{15}{5}$ as there are 2's in 15; there are $7\frac{1}{2}$ two's in 15; therefore $7\frac{1}{2}$ pecks can be bought.

13. When 1 peach is worth $\frac{3}{4}$ of an orange, how many peaches can be bought for 9 oranges?

14. How many squashes at $\$\frac{3}{8}$ apiece can be bought for \$6?

15. 6 are how many times $\frac{3}{8}$? $\frac{3}{11}$?

16. How many jars of preserves at $\$\frac{5}{6}$ a jar can be bought for \$10?

17. In how many days can Helen knit 2 scarfs, if she knits $\frac{4}{7}$ of a scarf in 1 day?

18. Mrs. Grant spent \$12 for Christmas wreaths at $\$\frac{3}{4}$ apiece; how many wreaths did she buy?

19. 10 are how many times $\frac{3}{4}$? $\frac{6}{7}$?

20. 5 are how many times $\frac{5}{9}$? $\frac{8}{9}$?

72. 1. Ella sold a lamp mat for $\$\frac{5}{6}$, and received in payment some cambric at $\$\frac{2}{5}$ a yard; how many yards did she receive?

Solution. She received as many yards as there are 2's of fifths in $\frac{5}{6}$. We first change $\frac{2}{5}$ and $\frac{5}{6}$ to fractional numbers having a common denominator. $\frac{2}{5}$ equals $\frac{12}{30}$, and $\frac{5}{6}$ equals $\frac{25}{30}$. There are as many 12's of 30ths in $\frac{25}{30}$ as there are 12's in 25, which is $2\frac{1}{12}$; therefore $2\frac{1}{12}$ yards can be bought.

2. At $\$\frac{2}{3}$ a bushel, how many bushels of apples can be bought for $\$\frac{5}{6}$?

3. At $\$\frac{4}{9}$ a pound, how many pounds of butter can be bought for $\$\frac{2}{3}$?

4. If an ounce of worsted costs $\$\frac{3}{10}$, how many ounces can be bought for $\$\frac{2}{5}$?

5. How many bottles of ink at $\$\frac{3}{20}$ a bottle can be bought for $\$\frac{3}{5}$?

6. If $\frac{2}{9}$ of a pound of cotton is required to make 1 yard of cloth, how many yards can be made out of $\frac{5}{6}$ of a pound? out of $1\frac{3}{5}$ pounds? out of $6\frac{2}{3}$?

NOTE. Change mixed numbers to fractional numbers.

7. If a milliner puts $\frac{4}{7}$ of a yard of wire into a hat, how many hats can she wire with $3\frac{1}{2}$ yards?

8. How many dominoes can be cut from a piece of ivory $6\frac{3}{4}$ inches long, if $1\frac{1}{2}$ inches in length is sufficient to make 1 domino?

9. How many checkers can be sawed from a piece of wood $1\frac{3}{4}$ inches long, allowing $\frac{3}{8}$ of an inch of the length for 1 checker?

10. How many dozen tomato plants at $\$\frac{5}{8}$ a dozen can be bought for $\$2\frac{1}{2}$?

11. If a wagon wheel is $3\frac{1}{3}$ yards in circumference, how many times will it turn in going a distance of $5\frac{5}{7}$ yards?

12. Jane had $\$1\frac{4}{5}$, and received $\$\frac{2}{5}$ more: how many yards of cambric at $\$\frac{2}{3}$ a yard could she buy with her money?

13. Granville received $\$18\frac{2}{3}$ for work; and after paying a debt of $\$12\frac{5}{12}$, he found that he had just money enough to pay for his board for $1\frac{1}{2}$ weeks; what was the price of his board for a week?

14. Mr. Jenks sold $2\frac{1}{2}$ bushels of corn at $\$\frac{2}{3}$ a bushel, and received his pay in butter at $\$\frac{2}{5}$ a pound; how many pounds of butter did he receive?

15. 15 are how many times $1\frac{1}{2}$? $4\frac{1}{2}$?

16. 4 are how many times $2\frac{1}{3}$? $4\frac{1}{9}$?

17. $\frac{5}{8}$ are how many times $\frac{2}{3}$? $\frac{3}{4}$?

18. $\frac{4}{11}$ are how many times $\frac{2}{5}$? $\frac{3}{8}$?

19. Divide $\frac{8}{10}$ by $\frac{1}{8}$; $\frac{5}{9}$ by $\frac{1}{5}$; $\frac{9}{11}$ by $\frac{7}{8}$.

20. Divide $\frac{4}{5}$ by $\frac{7}{10}$; $1\frac{2}{3}$ by $2\frac{1}{2}$; $4\frac{3}{5}$ by $1\frac{1}{2}$.

21. Divide $8\frac{1}{4}$ by $3\frac{2}{3}$; $1\frac{1}{9}$ by $2\frac{1}{2}$; $1\frac{4}{7}$ by $2\frac{1}{4}$.

Section XXXIV.

GENERAL REVIEW IN FRACTIONS.

73. 1. At $\frac{1}{2}$ of a dollar a day for board, how many days' board can be obtained for $7\frac{1}{2}$ dollars?

2. If a number of dollars is divided into 10 equal parts, and one of the parts is $\$\frac{1}{5}$, what is the number?

3. A traveller found, after paying $\$5\frac{1}{2}$ for passage money and $\$3\frac{1}{3}$ for refreshments, that he had $\$7\frac{2}{5}$ left; how much money had he at first?

4. A certain number increased by $6\frac{3}{4}$ equals $17\frac{1}{8}$; what is the number?

5. If 5 men can build a barn in $\frac{6}{7}$ of a week, in what time can 1 man build it?

6. A man bought a quantity of drugs for $53, and was obliged to sell them for $\frac{9}{10}$ of what they cost him; how much did he lose?

7. A man bought land at $\$11\frac{2}{3}$ an acre, and sold it at $\$19\frac{1}{8}$; what did he gain on 1 acre? on 3 acres?

8. What number added to $\frac{3}{4}$ of $10\frac{1}{3}$ will equal $12\frac{1}{2}$?

9. A miller paid $\$30\frac{1}{3}$ for some wheat, and sold it so as to gain $\frac{3}{4}$ as much as it cost; what did he receive for it?

10. If a man bought a watch for $\$24\frac{2}{3}$, and sold it for $\frac{4}{3}$ of what he gave, what did he gain?

11. Mr. Emery sold 12 cabbages at $8\frac{1}{3}$ cents apiece, and received in payment 6 pounds of meat at $11\frac{1}{2}$ cents a pound, and the rest in money; how much did he receive in money?

12. If 3 posts cost $\$4\frac{1}{2}$, what will 8 posts cost?

13. If 5 yards of silk cost $\$10\frac{1}{2}$, what cost 2 yards?

14. If 6 pounds of lead cost $\$1\frac{1}{7}$, what cost 5 pounds?

15. If 8 bottles of ink cost $\$2\frac{2}{5}$, what cost 5 bottles?

16. If 7 pounds of starch lasts a family $1\frac{1}{3}$ months, how long will 11 pounds last them?

17. If, when a man travels 8 hours a day, he can perform a journey in $10\frac{2}{3}$ days, in how many days would he perform it if he travelled 10 hours a day?

18. If 9 house lots of 1 acre each can be made from a certain lot of land, how many lots of $\frac{1}{4}$ of an acre each can be made from the same land? How many lots of $\frac{3}{4}$ of an acre each?

19. If a piece of cloth 1 yard wide weighs $1\frac{1}{10}$ pounds to the yard, what will the same kind of cloth weigh if it is $\frac{1}{2}$ of a yard wide? if it is $1\frac{1}{2}$ yards wide?

20. If a piece of cloth $\frac{3}{4}$ of a yard wide is worth $\$7\frac{1}{3}$, what is a piece of the same length, $\frac{5}{4}$ of a yard wide, worth?

21. If a staff 8 feet long casts a shadow of $\frac{2}{3}$ of a foot in length, what length of shadow would be cast at the same time of day by a staff 1 foot long? by a staff 20 feet long?

22. At \$4 a pound for ivory, how much can be bought for $\$1\frac{1}{6}$?

23. Mr. James has received \$12 for carrying a lot of goods; if he received $\$\frac{2}{3}$ for each load, how many loads did he carry?

24. How many balls of $\frac{1}{4}$ of a pound each are contained in a lot of balls weighing 15 pounds, the balls being of equal size?

25. How many rations of $1\frac{1}{4}$ pounds each can be furnished from 11 pounds of meat?

26. If, by working 10 hours a day, a person performs a piece of work in $2\frac{1}{5}$ days, in how many days can he perform it by working $7\frac{1}{2}$ hours a day?

27. How many bushels of peas at $\$2\frac{1}{2}$ a bushel can be bought for $\$5\frac{1}{4}$?

Section XXXV.

TO FIND THE WHOLE WHEN ONE OR MORE OF THE FRACTIONAL PARTS IS GIVEN. (See SECT. XIX.)

74. 1. If $\frac{1}{2}$ of a yard of gimp costs 3 cents, what will 1 yard cost at the same rate?

2. If $\frac{1}{3}$ of a pound of sugar is worth 4 cents, what is a whole pound worth?

3. If 2 apples will pay for $\frac{1}{4}$ of a melon, how many apples will pay for a whole melon?

4. In $\frac{1}{6}$ of a quire of paper there are 4 sheets; how many sheets are there in 1 quire?

5. Mr. Orne sold a boat and gained $8, which was $\frac{1}{7}$ of what it cost him; how much did it cost him?

6. 8 is $\frac{1}{7}$ of what number?

7. A man sold a carriage and lost $11, which was $\frac{1}{12}$ of what it cost him; what did it cost him?

8. A man sold a watch for $\frac{2}{3}$ of what it cost him, and thereby lost $3; what did the watch cost him?

9. 3 is $\frac{1}{3}$ of what number?

10. The age of Susan is $\frac{6}{5}$ of the age of Mary, and the difference between their ages is 2 years; what is the age of Mary?

Suggestion. Mary's age is $\frac{5}{5}$ of itself; the difference between $\frac{6}{5}$ and $\frac{5}{5}$ is $\frac{1}{5}$.

11. 2 is $\frac{1}{5}$ of what number?

12. A pole is standing so that $\frac{7}{8}$ of it is in the air and 4 feet in the earth; how long is the pole?

13. 4 is $\frac{1}{8}$ of what number? $\frac{1}{11}$ of what number?

14. After losing $\frac{2}{3}$ of his sheep, a farmer had 12 sheep left; how many sheep had he at first?

15. 12 is $\frac{1}{3}$ of what number? $\frac{1}{10}$ of what number?

16. $1\frac{1}{2}$ is $\frac{1}{4}$ of what number? $\frac{1}{7}$ of what number?

17. $8\frac{1}{3}$ is $\frac{1}{6}$ of what number? $\frac{1}{12}$ of what number?

18. $16\frac{2}{3}$ is $\frac{1}{2}$ of what number? $\frac{1}{3}$ of what number?

19. 20 is $\frac{1}{5}$ of what number? $\frac{1}{6}$ of what number?

75. 1. If $\frac{2}{5}$ of a piece of work can be performed in 4 days, in how many days can $\frac{1}{5}$ of the work be performed? In how many days can the whole work be performed?

2. 4 is $\frac{2}{5}$ of what number?

Solution. If 4 is $\frac{2}{5}$ of some number, $\frac{1}{5}$ of the number is $\frac{1}{2}$ of 4, which is 2; and $\frac{5}{5}$, or the whole number, is 5 times 2, or 10; therefore, etc.

3. A man performed $\frac{3}{4}$ of a journey in 15 days; in what time could he perform the whole journey at the same rate?

4. 15 is $\frac{3}{4}$ of what number?

5. If $\frac{5}{8}$ of a pound of figs costs 10 cents, what will 1 pound cost?

6. 10 is $\frac{5}{8}$ of what number?

7. In $\frac{7}{9}$ of a page there are 21 lines; how many lines are there in a whole page?

8. 21 is $\frac{7}{9}$ of what number?

9. A man sold a cow and gained \$20, which was $\frac{2}{7}$ of what she cost him; how much did she cost him? What did he receive for her?

10. From a cask of spirits, 15 gallons leaked out, which was $\frac{3}{10}$ of the contents of the cask; what were the contents of the cask at first?

11. 15 is $\frac{3}{10}$ of what number?

12. If an author's copyright is $\frac{2}{11}$ of the whole profit upon a book, and he receives 4 hundred dollars, what is the whole profit?

13. 4 is $\frac{2}{11}$ of what number?

14. A man gained \$15 by selling a cow for $1\frac{3}{4}$ times what she cost him; what did she cost him?

15. 15 is $\frac{3}{4}$ of what number?

16. Edgar sold a top for $\frac{7}{5}$ of what it cost him, and thereby gained 8 cents; what did the top cost him?

17. 8 is $\frac{2}{5}$ of what number?

18. A grocer sold cheese for $\frac{1}{5}$ of what it cost him, and thereby lost 8 cents on a pound; what did he give a pound?

19. 8 is $\frac{4}{5}$ of what number?

20. If $\frac{3}{8}$ of the sheep in a flock are sold, 55 sheep will remain; how many sheep are there in the flock?

21. 55 is $\frac{5}{8}$ of what number?

22. If after a man has spent $\frac{2}{11}$ of his money he has $36 remaining, how much money had he at first?

23. 36 is $\frac{9}{11}$ of what number?

24. If a man sells a firkin of butter for $16, and thereby gains $\frac{1}{3}$ of what it cost him, how much did it cost him?

25. 16 is $\frac{4}{3}$ of what number?

26. A watchmaker bought a watch and chain: he paid for the chain $18, which was $\frac{6}{11}$ of what he paid for the watch; what did he pay for the watch? What did he pay for both?

27. By selling beef at 8 cents a pound, a market-man loses $\frac{1}{3}$ of what his meat cost him; what did it cost him a pound?

28. 8 is $\frac{2}{3}$ of what number?

29. After spending $\frac{1}{10}$ of her money, a woman has $27 remaining; how much money had she at first?

30. 27 is $\frac{9}{10}$ of what number? $\frac{3}{8}$ of what number?

31. 24 is $\frac{8}{9}$ of what number? $\frac{3}{10}$ of what number?

32. 36 is $\frac{4}{5}$ of what number? $\frac{9}{8}$ of what number?

33. 45 is $\frac{9}{8}$ of what number? $\frac{5}{6}$ of what number?

34. 49 is $\frac{7}{11}$ of what number? $\frac{7}{10}$ of what number?

35. 64 is $\frac{8}{7}$ of what number? $\frac{8}{3}$ of what number?

36. 108 is $\frac{12}{5}$ of what number? $\frac{9}{4}$ of what number?

76. 1. If $\frac{3}{4}$ of a bale of cotton cost \$30, what will 1 bale cost at the same rate? How many yards of cloth at \$5 a yard will pay for 1 bale of cotton?

2. 30 is $\frac{3}{4}$ of how many times 5?

Solution. If 30 is $\frac{3}{4}$ of some number, $\frac{1}{4}$ of the number is $\frac{1}{3}$ of 30, which is 10; and $\frac{4}{4}$, or the whole number, is 4 times 10, or 40; and 40 is 8 times 5; therefore, 30 is $\frac{3}{4}$ of 8 times 5.

3. If $\frac{7}{8}$ of an acre of land is sold for \$21, what will a whole acre bring at the same rate? How many sheep at \$2 each will pay for 1 acre of the land?

4. 21 is $\frac{7}{8}$ of how many times 2?

5. I sold a horse for \$48, which is $\frac{8}{9}$ of what he cost me: if I paid for the horse with coal at \$10 a ton, how many tons did it take?

6. 48 is $\frac{8}{9}$ of how many times 10?

7. $\frac{4}{5}$ of a person's wages for a month are \$28; how many weeks' board at \$3 a week can he pay with his wages for 1 month?

8. 28 is $\frac{4}{5}$ of how many times 3?

9. 42 is $\frac{6}{7}$ of how many times 8?

10. 44 is $\frac{11}{3}$ of how many times 10?

11. 54 is $\frac{9}{11}$ of how many times 12?

12. 80 is $\frac{10}{7}$ of how many times 9?

13. 110 is $\frac{10}{11}$ of how many times 12?

14. 56 is $\frac{7}{4}$ of how many times 16?

77. 1. Jane had 24 cents and spent $\frac{3}{4}$ of her money for pencils at 9 cents apiece; how many pencils did she buy?

Solution. She bought as many pencils as there are times 9 in $\frac{3}{4}$ of 24: $\frac{1}{4}$ of 24 is 6, and $\frac{3}{4}$ of 24 is 3 times 6, or 18; and 18 is 2 times 9; therefore, she bought 2 pencils.

2. Mark had 42 apples, and gave $\frac{2}{7}$ of them for 4 fish-hooks; how many apples did he give for 1 fish-hook?

3. $\frac{2}{7}$ of 42 is how many times 4?

4. $\frac{5}{12}$ of 36 is how many times 5?

5. $\frac{4}{7}$ of 35 is how many times 6?

6. $\frac{3}{8}$ of 64 is how many times 7?

7. $\frac{6}{10}$ of 70 is how many times 8?

8. $\frac{9}{11}$ of 33 is how many times 9?

78. 1. Mary found 45 early violets, and $\frac{2}{9}$ of the number she found was equal to $\frac{5}{7}$ of the number which Hattie found; how many did Hattie find?

Solution. $\frac{1}{9}$ of 45 is 5, and $\frac{2}{9}$ are 2 times 5, or 10: if 10 is $\frac{5}{7}$ of some number, $\frac{1}{7}$ of the number must be $\frac{1}{5}$ of 10, which is 2, and $\frac{7}{7}$, or the whole number, must be 7 times 2, or 14; therefore, Hattie found 14 early violets.

2. $\frac{8}{5}$ of 25 is $\frac{4}{7}$ of what number?

3. $\frac{4}{9}$ of 18 is $\frac{2}{11}$ of what number?

4. $\frac{6}{13}$ of 26 is $\frac{4}{9}$ of what number?

5. $\frac{3}{4}$ of 28 is $\frac{7}{12}$ of what number?

6. $\frac{11}{12}$ of 48 is $\frac{4}{5}$ of what number?

79. 1. If a ton of hay is worth \$20, and if $\frac{3}{5}$ of a ton will pay for $\frac{2}{3}$ of a piece of cloth, how many barrels of apples at \$4 a barrel will pay for 1 piece of the cloth?

Solution. If a ton of hay is worth \$20, $\frac{3}{5}$ of a ton is worth $\frac{3}{5}$ of \$20, or \$12: if \$12 will pay for $\frac{2}{3}$ of a piece of cloth, $\frac{1}{2}$ of \$12, or \$6, will pay for $\frac{1}{3}$ of the cloth; and 3 times \$6, or \$18, will pay for the whole piece; but the cloth is to be paid for with apples at \$4 a barrel, and it will require as many barrels at \$4 a barrel as there are times \$4 in \$18, which is $4\frac{1}{2}$ times; therefore, $4\frac{1}{2}$ barrels of apples will pay for the cloth.

2. $\frac{5}{8}$ of 48 is $\frac{5}{6}$ of how many times 7?
3. $\frac{3}{7}$ of 35 is $\frac{5}{12}$ of how many times 9?
4. $\frac{3}{5}$ of 50 is $\frac{10}{11}$ of how many times 6?
5. $\frac{8}{9}$ of 63 is $\frac{7}{6}$ of how many times 12?
6. $\frac{2}{15}$ of 30 is $\frac{4}{13}$ of how many times 8?
7. $\frac{16}{20}$ of 40 is $\frac{8}{9}$ of how many times 6?

80. 1. If John is 16 years of age, and $\frac{3}{4}$ of his age is equal to $\frac{6}{7}$ of 4 times Willie's age, what is Willie's age?

Solution. If John's age is 16 years, $\frac{3}{4}$ of his age is 12 years: if 12 years is $\frac{6}{7}$ of 4 times Willie's age, $\frac{1}{7}$ of 4 times Willie's age is $\frac{1}{6}$ of 12 years, or 2 years; and $\frac{7}{7}$ of 4 times Willie's age is 7 times 2 years, or 14 years; 14 years is 4 times $3\frac{1}{2}$ years; therefore, Willie's age is $3\frac{1}{2}$ years.

2. $\frac{3}{4}$ of 32 is $\frac{6}{7}$ of 5 times what number?
3. $\frac{5}{7}$ of 42 is $\frac{3}{4}$ of 8 times what number?
4. $\frac{6}{5}$ of 10 is $\frac{2}{9}$ of 10 times what number?

81. 1. Susan has 16 cents: $\frac{5}{4}$ of Susan's money equals $\frac{10}{3}$ of Mary's; now, if John has $\frac{1}{5}$ of a dime how many times as much money has Mary as John? (1 dime = 10 cents.)

Solution. If Susan has 16 cents, $\frac{5}{4}$ of Susan's money is $\frac{5}{4}$ of 16 cents, or 20 cents; if 20 cents is $\frac{10}{3}$ of Mary's money, $\frac{1}{3}$ of Mary's money must be $\frac{1}{10}$ of 20 cents, or 2 cents; and $\frac{3}{3}$ of her money must be 3 times 2 cents, or 6 cents; $\frac{1}{5}$ of 10 cents is 2 cents, and 2 cents is contained in 6 cents 3 times; therefore, Mary has 3 times as much money as John.

2. $\frac{2}{15}$ of 45 is $\frac{3}{7}$ of how many thirds of 33?
3. $\frac{5}{12}$ of 48 is $\frac{2}{7}$ of how many sixths of 36?
4. $\frac{7}{8}$ of 32 is $\frac{14}{15}$ of how many eighths of 24?

5. $\frac{9}{10}$ of 40 is $\frac{6}{11}$ of how many ninths of 54?
6. $\frac{10}{11}$ of 55 is $\frac{2}{3}$ of how many halves of 16?
7. $\frac{11}{12}$ of 96 is $\frac{8}{3}$ of how many sevenths of 63?

82. 1. Mary is $8\frac{2}{5}$ years old, which is $\frac{6}{7}$ of the age of John; how old is John?
2. $8\frac{2}{5}$ is $\frac{6}{7}$ of what number?
3. Grace has read $22\frac{1}{2}$ pages of her Botany, which is $\frac{5}{12}$ of the whole book; how many pages are there in the book?
4. $22\frac{1}{2}$ is $\frac{5}{12}$ of what number?
5. $2\frac{1}{7}$ is $\frac{5}{7}$ of what number?
6. $37\frac{1}{2}$ is $\frac{3}{8}$ of what number?
7. $49\frac{2}{7}$ is $\frac{5}{8}$ of what number?
8. $62\frac{1}{2}$ is $\frac{5}{7}$ of what number?
9. $66\frac{2}{3}$ is $\frac{2}{3}$ of what number?
10. $83\frac{1}{3}$ is $\frac{10}{12}$ of what number?
11. $\frac{2}{5}$ is $\frac{7}{8}$ of what number?
12. $\frac{8}{11}$ is $\frac{4}{13}$ of what number?
13. $1\frac{9}{11}$ is $\frac{5}{7}$ of what number?
14. $\frac{3}{4}$ is $\frac{5}{9}$ of what number?

83. 1. Mary had $\$\frac{1}{2}$, and paid $\frac{3}{4}$ of it for $\frac{2}{5}$ of a yard of cloth; what was the cloth a yard?
2. $\frac{3}{4}$ of $\frac{1}{2}$ is $\frac{2}{5}$ of what number?
3. Mrs. Doane had $\$3\frac{1}{2}$, and spent $\frac{3}{7}$ of the money for $\frac{2}{3}$ of a bushel of cranberries; what was the price of the cranberries a bushel?
4. $\frac{3}{7}$ of $3\frac{1}{2}$ is $\frac{2}{3}$ of what number?
5. $\frac{5}{8}$ of $\frac{4}{5}$ is $\frac{3}{10}$ of what number?
6. $\frac{8}{9}$ of $1\frac{2}{7}$ is $\frac{4}{5}$ of what number?
7. $\frac{9}{11}$ of $9\frac{1}{6}$ is $\frac{3}{12}$ of what number?

84. 1. If $\frac{1}{3}$ of a yard of silk costs \$$1\frac{1}{2}$, what will 10 yards cost?

2. If $\frac{1}{6}$ of a rod of wall can be made in $2\frac{1}{3}$ hours, in what time can 5 rods be made?

3. If $\frac{1}{4}$ of a barrel of beef will last a family 3 months, how long will $2\frac{1}{2}$ barrels last the family?

4. If $\frac{3}{4}$ of a cord of wood costs \$6, what will 7 cords cost?

5. Mr. King sold 2 calves for \$24, which was $\frac{4}{3}$ of what he paid for them; what did each of his calves cost him?

6. A farmer paid \$$5\frac{5}{8}$ for a sheep, which was $\frac{3}{7}$ of what he received for a calf; what did he receive for the calf?

7. If a man can walk $\frac{3}{7}$ of a mile in $7\frac{1}{2}$ minutes, in what time can he walk 5 miles?

8. The distance from Boston to Reading is $\frac{3}{8}$ of the distance from Boston to Bradford by the same route; if the distance from Reading to Bradford is 20 miles, what is the distance from Boston to Bradford?

9. A child is $2\frac{1}{2}$ years old: if he is $\frac{1}{10}$ as old as his mother, and his mother is $\frac{5}{6}$ as old as his father, how old is his father?

10. There is a barber's pole, 4 feet of which is painted white, 8 feet is striped, and the remainder, which is $\frac{1}{4}$ of the whole pole, is in the ground; what is the length of the pole?

11. After spending $\frac{5}{6}$ of his money, and giving away $\frac{1}{12}$ of it, Mr. Aldrich has \$9 left; how many dollars had he at first?

12. A man has mowed $\frac{5}{8}$ of a lot of land, and a boy has mowed $\frac{1}{3}$ of it: if the man can mow the rest of it in 2 hours, in what time could he have mowed the whole?

13. $\frac{1}{3}$ of my property is in my store, $\frac{1}{2}$ of the remainder is in the bank, and the balance, which is \$4000, is loaned to a friend; how much property have I?

14. $\frac{2}{3}$ of a certain number exceeds $\frac{1}{2}$ of it by $12\frac{1}{2}$; what is the number?

15. $\frac{3}{4}$ of a man's age exceeds $\frac{2}{3}$ of it by $4\frac{1}{2}$ years; what is his age?

16. If 20 cents are paid for $\frac{5}{6}$ of a melon, what will $\frac{2}{5}$ of a melon cost?

17. If the cost of sawing $\frac{3}{8}$ of a cord of wood is 33 cents, what is the cost of sawing $\frac{3}{10}$ of a cord?

18. If $\frac{6}{7}$ of a barrel of oil costs $36, what will $\frac{5}{6}$ of a barrel cost?

19. In $\frac{2}{3}$ of $\frac{3}{4}$ of a firkin of butter there are 21 pounds; how many pounds are there in $\frac{5}{7}$ of a firkin?

20. A man sells vinegar at 20 cents a gallon, which is $\frac{5}{3}$ of what it cost him; what did it cost him? What part of the cost was the gain?

21. If a tailor sold a coat for $25, and gained $\frac{1}{4}$ of what it cost, what did it cost?

22. A butcher has a cow and a calf, both of which he values at $30: estimating the value of the calf at $\frac{1}{5}$ of the value of the cow, what is the value of each?

Solution. $30 contains the value of the cow plus $\frac{1}{5}$ of her value, or $\frac{6}{5}$ of her value; if $30 is $\frac{6}{5}$ of her value, $\frac{1}{5}$ of her value must be $\frac{1}{6}$ of $30, etc.

23. If a hat and vest together cost $8, and the hat costs $\frac{1}{3}$ as much as the vest, what is the cost of each?

24. What number is that to which if $\frac{2}{7}$ of itself is added, the sum will equal 18?

25. What number is that of which if $\frac{2}{7}$ of itself be taken away, the remainder will equal 15?

26. If $\frac{2}{3}$ of a box of raisins is worth $4, how many clocks at $2 apiece will pay for the box of raisins?

27. When 1 ton of coal is worth $9, and $\frac{2}{3}$ of a ton of coal will pay for $\frac{3}{4}$ of a load of wood, how many pairs of boots, at $24 a dozen pairs, will be required to pay for a load of wood?

Section XXXVI.

CURRENCY.

UNITED STATES MONEY.

85.

TABLE.

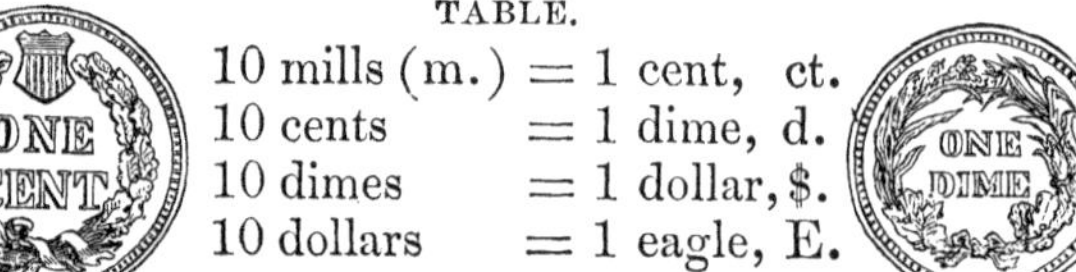

10 mills (m.) = 1 cent, ct.
10 cents = 1 dime, d.
10 dimes = 1 dollar, \$.
10 dollars = 1 eagle, E.

1. How many mills in 2 cents? in 3 c. 7 m.?
2. How many cents in 3 dimes? in 3 d. 9c.?
3. How many cents in 1 dollar? in \$2? in \$5?
4. How many mills in 1 dollar? in \$2? in \$4?
5. How many dollars in 150 cents, and what remains?
6. How many dollars in 283 cents, and what remains?
7. How many cents in 20 mills? in 100 mills? in 150 mills?

In business operations, the denominations eagle and dime are not used; eagles are regarded as tens of dollars, and dimes as tens of cents; thus, 2 E., \$3, 4 d., 5 c., are called 23 dollars and 45 cents, and are written \$23.45.

The sign for dollars (\$) is placed before figures which represent United States Money, and a dot or period is placed between the figures which represent dollars and those which represent cents.

Mills are represented by figures in the third place at the right of the period.

Read the following: —

8. \$53.27; \$28.02; \$37.053; \$10.001.
9. \$19.20; \$30.94; \$82.002; \$2.192.

Write the following : —

10. 2 dollars, 25 cents, 4 mills.
11. 5 dollars, 28 cents, 7 mills.
12. 31 dollars, 16 cents, 2 mills.
13. 83 dollars, 15 cents, 6 mills.
14. 27 dollars, 2 cents.
15. 30 dollars, 3 mills.
16. How many cents in \$1? in $\$\frac{1}{2}$? $\$\frac{1}{4}$? $\$\frac{3}{4}$?
17. How many cents in $\$\frac{1}{8}$? in $\$\frac{3}{8}$? $\$\frac{5}{8}$? $\$\frac{7}{8}$?
18. How many cents in $\$\frac{1}{3}$? in $\$\frac{2}{3}$? $\$\frac{1}{6}$? $\$\frac{5}{6}$?
19. How many cents in $\$\frac{1}{5}$? in $\$\frac{2}{5}$? $\$\frac{3}{5}$? $\$\frac{4}{5}$?
20. How many cents in $\$\frac{1}{10}$? in $\$\frac{3}{10}$? $\$\frac{7}{10}$ $\$\frac{9}{10}$?
21. What part of \$1 is 10 cents? $12\frac{1}{2}$ cts.? $16\frac{2}{3}$ cts.? 20 cts.? 25 cts.? 30 cts.? $33\frac{1}{3}$ cts.? $37\frac{1}{2}$ cts.?
22. What part of \$1 is 40 cts.? 50 cts.? 60 cts.? $62\frac{1}{2}$ cts.? 70 cts.? 75 cts.? $83\frac{1}{3}$ cts.? $87\frac{1}{2}$ cts.?

86. A **Coin** is a piece of metal having a fixed value attached to it by Government.

The United States coins are made of gold, silver, copper and nickel.

The gold coins are the		The silver coins are the	
Double-eagle, . .	\$20.00	Dollar, ...	\$1.00
Eagle, ...	10.00	Half-dollar, ...	.50
Half-eagle, . . .	5.00	Quarter-dollar, . .	.25
Quarter-eagle, . .	2.50	Dime, ...	.10
Three-dollar piece,	3.00	Half-dime, ...	.05
One-dollar piece, .	1.00	Three-cent piece, .	.03

The copper and nickel pieces are one-cent, two-cent, three-cent, and five-cent pieces.

Besides these coins, paper currency is used, which has the same representative value as the coins.

1. A gentleman bought a hat for 1 half of a dollar, and sold it for 75 cents; how many cents did he gain?

2. Jane had $\$\frac{4}{5}$, and George had 87 cents; how much more money had George than Jane?

3. Alfred had $\$\frac{9}{10}$, and spent 28 cents for a book, 25 for a knife, and 7 for a string: the money he had left was in 2 pieces; what must have been the value of each piece?

4. Annie had $1.25, and bought 3 yards of cambric at $12\frac{1}{2}$ cents a yard, some buttons for $12\frac{1}{2}$ cents, and some hooks and eyes for 4 cents: she gave in payment $1, and the merchant gave her the change in the smallest number of pieces possible; what was the value of each piece? How much money had she then?

5. Charles had two 25-cent pieces, three 10-cent pieces, six 3-cent pieces, and eleven single cents; these he exchanged for the same value in the smallest number of pieces possible; what pieces did he receive in exchange?

87. ENGLISH MONEY.

TABLE.

4 farthings (qr.)	= 1 penny,	d.
12 pence	= 1 shilling,	s.
20 shillings	= 1 pound,	£.

1. How many farthings in 2 pence? in 1 s. or 12 d.?

2. How many pence in 1 shilling? in 1 s. 6 d.?

3. How many shillings in 1 pound? in 2 £ 7 s.?.

4. In 50 shillings how many pounds, and what remains?

5. In 37 pence how many shillings, and what remains?

6. In 22 farthings how many pence, and what remains?

Section XXXVII.

WEIGHT.

88. Avoirdupois Weight.

This weight is used for weighing almost all articles except gold, silver, and precious stones.

TABLE.

16 ounces (oz.)	= 1 pound, ...	lb.
100 pounds	= 1 hundred weight,	cwt.
20 hundred weight	= 1 ton, ...	T.

Note. By the above scale of weights the ton contains 2000 pounds; the *long* ton of 2240 pounds is sometimes used for weighing gross articles, as iron and coal, and is the ton recognized by the United States Government. By this scale the hundred weight equals 112 pounds, and is divided into 4 quarters of 28 pounds each.

1. How many ounces in 2 pounds? in 3 lbs. 2 oz.?
2. How many pounds in 1 cwt.? how many oz.?
3. How many cwt. in 1 ton? how many lbs.?
4. How many pounds in 20 ounces? in 40 oz.?
5. How many cwt. in 200 lbs.? in 1000 lbs.?
6. How many tons in 40 cwt.? in 90 cwt.?
7. What part of 1 pound is 8 oz.? 6 oz.? 15 oz.?
8. What part of 100 weight is 20 lbs.? 36 lbs.?
9. What part of 1 ton is 2 cwt.? 10 cwt.? 15 cwt.?
10. How many oz. in $\frac{1}{2}$ lb.? $\frac{1}{4}$ lb.? $\frac{1}{6}$ lb.? $\frac{7}{8}$ lb.?
11. How many lbs. in $\frac{3}{5}$ cwt.? $\frac{1}{2}$ cwt.? $2\frac{1}{2}$ cwt.?
12. If 1 cwt. of flour costs $6, what is the cost of 1 lb.? of 50 lbs.?
13. If I buy 1 cwt. of sugar for $10, and sell it at 14 cents a pound, how much money do I gain on 1 pound? on 1 cwt.?

89. TROY WEIGHT.

This weight is used for weighing gold, silver, and precious stones, and where great accuracy is required.

TABLE.

24 grains (gr.)	=	1 pennyweight, pwt.
20 pennyweights	=	1 ounce, . . . oz.
12 ounces	=	1 pound, . . . lb.

1. How many grains in 1 pwt.? in 2 pwt. 3 gr.?
2. How many pwts. in 3 oz.? in 4 oz. 2 pwt.?
3. How many oz. in 5 lbs.? in 7 lbs. 8 oz.?
4. How many lbs. in 25 oz.? in 39 oz.?
5. How many oz. in 65 pwt.? in 82 pwt.?

90. APOTHECARIES' WEIGHT.

This weight is used by apothecaries for mixing medicines.

TABLE.

20 grains (gr.)	=	1 scruple, sc. or ℈.
3 scruples	=	1 dram, dr. or ʒ.
8 drams	=	1 ounce, oz. or ℥.
12 ounces	=	1 pound, lb. or ℔.

1. How many grains in 2 scruples? in 1 sc. 15 gr.?
2. How many scruples in 4 drams? in 5 dr. 2 sc.?
3. How many drams in 8 ounces? in 1 pound?
4. How many ounces in 2 pounds? in 2 lb. 4 oz.?
5. Change 100 grains to units of higher denominations. *Ans.* 1 dr. 2 sc.
6. Change 100 drams to units of higher denominations.

NOTE. Pounds, ounces, and grains of Troy and Apothecaries' weights are alike. A pound Avoirdupois is heavier than a pound Troy or Apothecaries' weight; but an ounce Avoirdupois is lighter than an ounce of the other weights.

Section XXXVIII.

EXTENSION.

91. LONG MEASURE.

TABLE.

12 inches (in.)	= 1 foot,	.	ft.
3 feet	= 1 yard,	.	yd.
$5\frac{1}{2}$ yards, or $16\frac{1}{2}$ feet	= 1 rod, .	.	rd.
320 rods	= 1 mile,	.	m.

One inch.

1. Change 1 yard 1 foot to feet; to inches.
2. Change 2 rods to yards; 3 rd. 3 yd. to yards.
3. Change 1 mile to yards; 1 mile to feet.
4. Change 100 in. to units of higher denominations.
5. Change 22 yards to units of higher denominations.
6. Change 90 rods to units of a higher denomination.
7. Change 640 rods to units of a higher denomination.
8. Change $\frac{1}{2}$ of a yard to units of lower denominations.

Solution. $\frac{1}{2}$ of a yard $=\frac{1}{2}$ of 3 ft. $=1\frac{1}{2}$ ft. $\frac{1}{2}$ ft. $=\frac{1}{2}$ of 12 in. $=6$ in. *Ans.* 1 ft. 6 in.

9. Change $\frac{1}{4}$ of a yard to units of lower denominations.
10. Change $\frac{1}{2}$ of a rod to units of lower denominations; $\frac{1}{10}$ of a mile to units of lower denominations.
11. What part of 1 mile is 1 rod? are 4 rods?
12. What part of 1 rod is 1 yard?

NOTE. Change 1 rod and 1 yard to half yards.

13. What part of 1 rod are 2 yards? $2\frac{1}{2}$ yards?

NOTE. Each pupil should procure, or make for his own use, measures of the foot marked off into inches, of the yard and the rod. A piece of tape can be made to answer the purpose.

8

14. Draw upon your slate or blackboard a line that you think is 1 inch long; 3 inches long; 10 inches long; 1 foot long.

15. Mark off in some convenient place a distance of 3 ft.; 6 ft.; 5 ft.

16. When out of school, measure 1 rod upon the ground, and count your steps in walking that distance.

17. Knowing the number of steps you take in walking the distance of 1 rod, estimate the length of the school-yard, the width of the street, the distance to your home, etc.

92. SQUARE MEASURE.

This measure is used for finding the area of surfaces, or whatever has length and breadth.

TABLE.

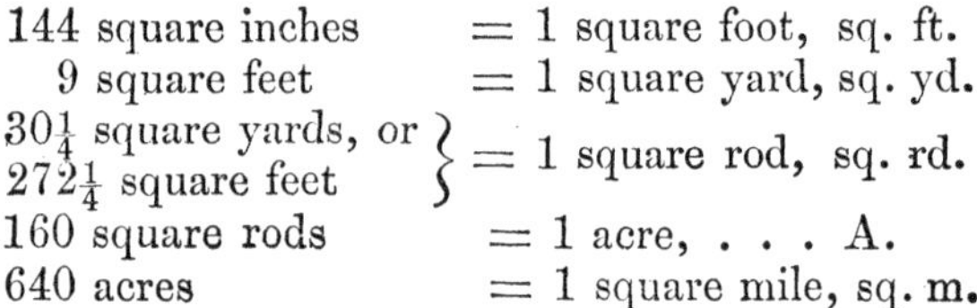

144 square inches	=	1 square foot, sq. ft.
9 square feet	=	1 square yard, sq. yd.
$30\frac{1}{4}$ square yards, or $272\frac{1}{4}$ square feet	=	1 square rod, sq. rd.
160 square rods	=	1 acre, . . . A.
640 acres	=	1 square mile, sq. m.

93. A Square is a four-sided figure which has all its sides equal, and all its corners or angles equal. A square 1 inch long and 1 inch wide, contains 1 square inch; a square 1 ft. long and 1 ft. wide, contains 1 sq. ft.; a square 1 yd. long and 1 yd. wide, contains 1 sq. yd.; a square 1 rd. long and 1 rd. wide, contains 1 sq. rd.; a square 1 mile long and 1 mile wide, contains 1 sq. m.

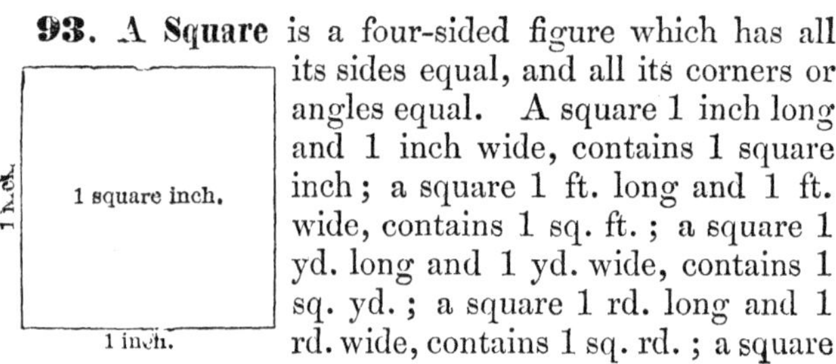

A **Rectangle** is a four-sided figure which has its opposite sides equal, and all its angles equal.

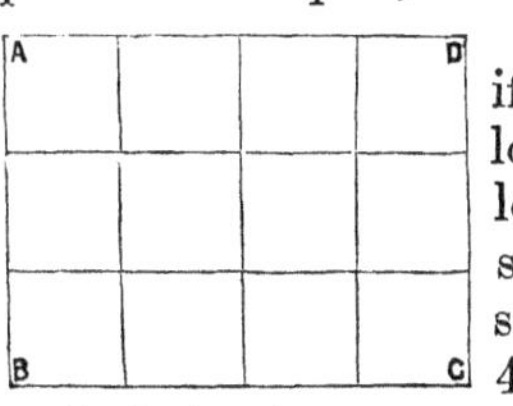

In the accompanying figure, if B C represents a line 4 inches long, and A B a line 3 inches long, the whole represents a surface containing 3 rows of 4 square inches each, which equals $4 \times 3 = 12$ square inches; hence, to find the **Area**, or contents of a rectangle, *Multiply the number of units in its length by the number of like units in its breadth.*

1. How long must a piece of cloth be which is 1 yard wide, that it may contain 2 square yards? 3 sq. yds.?

2. How many square yards of surface can be covered by 8 yards of carpeting which is 1 yard wide? by 5 yards which is 2 yards wide? by 12 yards which is $\frac{1}{2}$ a yard wide?

3. How many square feet are there in a table-cover which is 6 feet long and 6 feet wide? How many square yards are there in it?

4. How many yards of carpeting 1 yard wide will be required to carpet a hall that is 12 feet long and 6 feet wide?

5. How many square yards of carpeting will be required to carpet a room 12 feet square, that is, measuring 12 feet each way?

6. What is the difference between 2 feet square and 2 square feet? 9 inches square and 9 square inches?

7. How many square yards in a room 9 feet square?

8. How many square feet in the surface of a board that is 10 feet long and 8 inches wide?

NOTE. In this and in all similar examples, the given dimensions must be reduced to the same denomination before multiplying; thus, 8 in. $= \frac{2}{3}$ ft. $\frac{2}{3} \times 10 = \frac{20}{3} = 6\frac{2}{3}$. *Ans.* $6\frac{2}{3}$ feet.

9. How many square feet in a board 8 ft. long and 18 in. wide?

10. How many square yards in a roof 5 yds. long and $3\frac{1}{2}$ yds. wide?

11. How many square inches in a slate 8 in. long and $6\frac{1}{2}$ in. wide?

12. If a field which contains 60 square rods is 12 rods in length, what is its width?

94. CUBIC MEASURE.

This measure is used for finding the contents of anything that has length, breadth, and thickness.

TABLE.

1728 cubic inches (cu. in.)	= 1 cubic foot,	cu. ft.
27 cubic feet	= 1 cubic yard,	cu. yd.

95. In measuring firewood, and some other merchandise, the denomination *cord* is used. A pile of wood 4 ft. wide, 4 ft. high, and 8 ft. long, or its equivalent, contains 1 cord. A pile 4 ft. wide, 4 ft. high, and 1 ft. long, contains 1 cord foot. Hence,

16 cubic feet	= 1 cord foot,	cd. ft.
8 cord feet, or 128 cubic feet	= 1 cord, . .	cd.

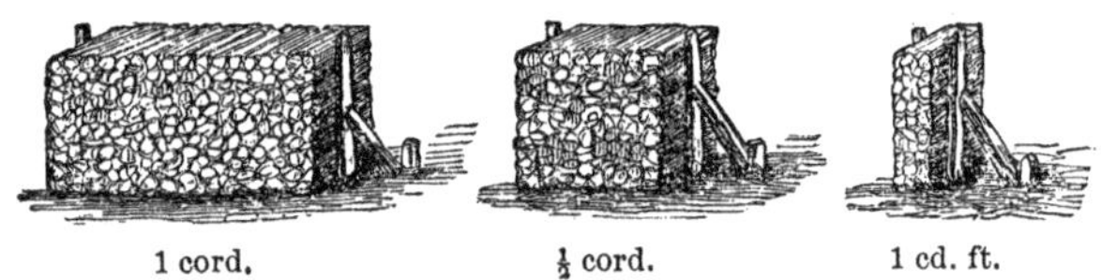

1 cord. $\frac{1}{2}$ cord. 1 cd. ft.

1. How many cords of wood in a pile 4 feet wide, 4 feet high, and 16 feet long? 20 ft. long? 32 ft. long?

2. How many cord feet in 1 cd. 7 cd. ft.?

3. If 1 cord of wood costs $10, what will $\frac{1}{2}$ cd. cost? 1 cd. ft.? 3 cd. ft.?

96. A **Cube** is a solid bounded by 6 equal squares.

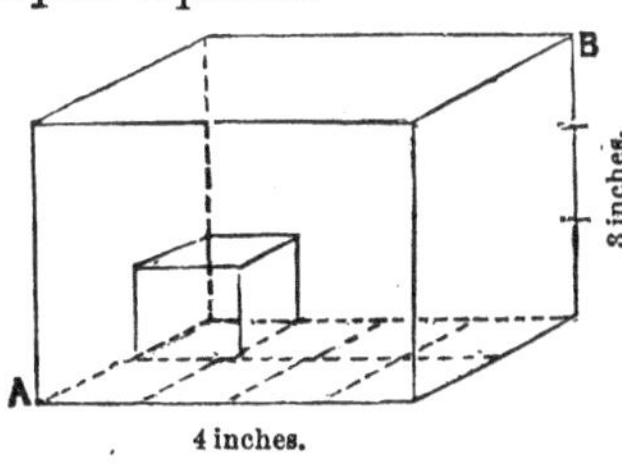

The accompanying figure represents a solid, 4 inches long, 2 inches wide, and 3 in. high; if the solid were 1 inch high, there would be standing upon its lower surface 4 × 2, or 8 cubic inches; but the solid is 3 inches high, and must therefore contain three 8's of cubic inches, or 24 cubic inches; hence, to find the solid contents of a cube, or of any solid bounded by rectangles, *Multiply the product of the number of units in its length and breadth by the number of like units in the height.*

1. How many cubic feet in a block 3 feet long, 1 foot wide, and 2 feet high?

2. How many cubic inches in a block 8 inches long, 3 inches wide, and 2 inches thick?

3. How many cubic inches does a box contain that measures on the inside 3 inches in width, 4 inches in depth, and 6 inches in length?

4. How much space is required to contain a box that measures on the outside 6 feet in length, 4 in breadth, and 3 in height?

5. How many square inches in the surface of a cubical block whose edges are 3 inches long?

6. How many cubic inches does the block in the above example contain?

7. How many square feet of carpeting will cover the top and sides of a box 4 feet long, 3 feet wide, and 2 feet high? How many square yards will be required?

Section XXXIX.

CAPACITY.

97. DRY MEASURE.

TABLE.

2 pints	=	quart,	qt.
8 quarts	=	1 peck,	pk.
4 pecks	=	1 bushel,	bu.

1. In 1 bushel how many pecks? how many quarts?
2. In 1 bu. 2 pk., how many pecks? quarts?
3. In 2 pk. 3 qt., how many quarts? pints?
4. Reduce 72 pints to units of higher denominations.
5. If a man bought 3 pecks of peas of one person, 2 pecks of another, and 5 pecks of another, how many bushels did he buy?
6. Mr. Green had 1 bushel of cranberries, and sold 7 quarts, how many quarts had he left?

98. LIQUID MEASURE.

TABLE.

4 gills (gi.)	=	1 pint,	pt.
2 pints	=	1 quart,	qt.
4 quarts	=	1 gallon,	gal.

1. In 4 quarts 1 pint, how many pints?
2. In 2 gal. 2 qt. 1 pt., how many pints?
3. Reduce 63 gills to units of higher denominations.
4. Reduce 77 gills to units of higher denominations.
5. Reduce $\frac{5}{6}$ of a quart to units of lower denominations.
6. Reduce $\frac{2}{3}$ of a gallon to units of lower denominations.

Section XL.

99. TIME MEASURE.

TABLE.

60 seconds (s.)	= 1 minute,	m. or min.
60 minutes	= 1 hour, .	h.
24 hours	= 1 day, .	d.
7 days	= 1 week, .	w.
365 days	= 1 common year,	c.y.
366 days	= 1 leap year, . .	l.y.
100 years	= 1 century, . .	C.

1. How many seconds in $\frac{1}{2}$ min.? in $\frac{1}{4}$ min.?
2. How many minutes in $\frac{1}{3}$ h.? in $\frac{1}{4}$ h.? $\frac{1}{6}$ h.?
3. How many hours in $\frac{1}{4}$ d.? in $1\frac{3}{4}$ d.?
4. How many days in 4 w.? in 6 w. 3 d.?
5. What part of 1 hour is 30 min.? 15 min.? 45 min.? 10 min.?

The year is divided into 12 months, as illustrated in the above diagram.

6. Name the months in their order, beginning with January.

7. Name the months that have 30 days in each.

8. Name the months that have 31 days in each.

9. How many days has February?

NOTE. In business operations a month is generally considered 30 days.

10. What part of 2 months, or 60 days, is 30 d.? 15 d.? 12 d.? 20 d.? 6 d.? 4 d.? 5 d.? 3 d.? 1 d.? 45 d.? 40 d.? 10 d.?

11. If there are 12 months in 1 year, how many months are there in 2 y.? in 3 y. 1 mo.?

12. What part of 200 months is 8 y. 4 mo.? 4 y. 2 mo.? 2 y. 1 mo.? 3 y. 4 mo.?

13. How many months from Jan. 1st to July 1st? from April 1st to Oct. 1st? from July 1st to Jan. 1st? from Oct. 1st to Jan. 1st? from Feb. 1st to May 1st? from Aug. 1st to Oct. 1st? from Mar. 1st to Jan. 1st? from Sept. 1st to July 1st?

14. In a leap year how many days are there from Jan. 1st to Apr. 1st? from Apr. 1st to July 1st? from July 1st to Oct. 1st? from Oct. 1st to Jan. 1st?

100. MISCELLANEOUS TABLES.

NUMBER.

12 units = 1 dozen.	12 gross = 1 great gross.
12 dozen = 1 gross.	20 units = 1 score.

PAPER.

24 sheets = 1 quire.	20 quires = 1 ream.

1. How many things in $\frac{1}{2}$ of a gross? $1\frac{1}{2}$ dozen?

2. How many years in 3 score years and 10? in 4 score?

3. At 1 cent a sheet, what will be the cost of 1 quire of paper? of 2 quires? of 1 ream?

Section XLI.

PERCENTAGE.

101. What is $\frac{5}{100}$ of 200 sheep? *Ans.* 10 sheep.

A number found by taking a number of hundreths of a given number, as 10 sheep above, is **Percentage.**

The number of hundredths which are taken to find a percentage, as 5 above, is the **rate per cent.**

100 per cent. $=\frac{100}{100}=1$	50 per cent. $=\frac{50}{100}=\frac{1}{2}$.
90 per cent. $=\frac{90}{100}=\frac{9}{10}$.	25 per cent. $=\frac{25}{100}=\frac{1}{4}$.
80 per cent. $=\frac{80}{100}=\frac{4}{5}$.	75 per cent. $=\frac{75}{100}=\frac{3}{4}$.
60 per cent. $=\frac{60}{100}=\frac{3}{5}$.	$37\frac{1}{2}$ per cent. $=\frac{37\frac{1}{2}}{100}=\frac{3}{8}$.
40 per cent. $=\frac{40}{100}=\frac{2}{5}$.	$12\frac{1}{2}$ per cent. $=\frac{12\frac{1}{2}}{100}=\frac{1}{8}$.
20 per cent. $=\frac{20}{100}=\frac{1}{5}$.	$6\frac{1}{4}$ per cent. $=\frac{6\frac{1}{4}}{100}=\frac{1}{16}$.
10 per cent. $=\frac{10}{100}=\frac{1}{10}$.	$66\frac{2}{3}$ per cent. $=\frac{66\frac{2}{3}}{100}=\frac{2}{3}$.
5 per cent. $=\frac{5}{100}=\frac{1}{20}$.	$33\frac{1}{3}$ per cent. $=\frac{33\frac{1}{3}}{100}=\frac{1}{3}$.
4 per cent. $=\frac{4}{100}=\frac{1}{25}$.	$16\frac{2}{3}$ per cent. $=\frac{16\frac{2}{3}}{100}=\frac{1}{6}$.
2 per cent. $=\frac{2}{100}=\frac{1}{50}$.	$8\frac{1}{3}$ per cent. $=\frac{8\frac{1}{3}}{100}=\frac{1}{12}$.

102. 1. What part of a number is 70 per cent. of the number? *Ans* $\frac{70}{100}$, which equals $\frac{7}{10}$ of the number.

2. What part of a number is 30 per cent. of it? 6 per cent. of it? $33\frac{1}{3}$ per cent. of it? $66\frac{2}{3}$ per cent. of it?

3. A stabler had 40 horses, and sold 10 per cent. of them; how many did he sell?

Ans. He sold $\frac{10}{100}$, or $\frac{1}{10}$ of 40 horses, which is 4 horses.

4. In a company of 60 persons, 20 per cent. were Scotch; how many persons were Scotch? 60 per cent. were English; how many were English?

5. A grocer had 12 barrels of flour, but he lost $16\frac{2}{3}$ per cent. of it by fire; how many barrels did he lose?

6. What is 5 per cent. of 20 trees? of 40 feet?

7. What is 90 per cent. of 60 men? of 75 days?

8. What is 1 per cent. of 100 weeks? of 200 years?

9. What is $1\frac{1}{2}$ per cent. of $200? of $300?

10. What is 2 per cent. of $400? of $220?

11. What is $2\frac{1}{4}$ per cent. of $200? $400? $1000?

12. What is 110 per cent. of 50 tons?

Solution. 100 per cent. of 50 tons is 50 tons; 10 per cent. of 50 tons equals $\frac{1}{10}$ of 50 tons, or 5 tons, which, added to 50 tons, equals 55 tons; therefore, etc.

13. What is 105 per cent. of 600 yards? of $40?

14. What is 120 per cent. of 200 miles? of $$62\frac{1}{2}$?

15. A trader bought a house-lot for $80, and sold it so as to gain 25 per cent.; what did he receive for it?

Suggestion. 25 per cent. of $80 equals $\frac{1}{4}$ of $80, or $20, which, added to $80, equals $100.

16. A man paid 35 cents a yard for cloth; what must he sell it for to gain 20 per cent.?

17. A merchant has tea which cost him 25 cents a pound: if he sells it so as to gain 100 per cent., what will he receive a pound for it?

18. A broker exchanged $16 in gold and gained 50 per cent. on the cost of it; what did he receive for it?

19. A landlord has advanced his rents 30 per cent.; what will be the rent of a tenement this year which rented last year at $20 a month?

20. At what price must oats be sold to gain $33\frac{1}{3}$ per cent., if they cost 30 cents a bushel? 36 cents a bushel?

103. The **Complement** of any given per cent. is the difference between the given per cent. and 100 per cent.; thus, the complement of 90 per cent. is 10 per cent.; the complement of 40 per cent. is 60 per cent.

1. What is the complement of 80 per cent.? of 75 per cent.? of 35 per cent.? of $12\frac{1}{2}$ per cent.?

2. A man who had 80 geese, lost 30 per cent. of them, and sold the remainder; what per cent. of them did he sell? how many geese?

3. From a cask containing 30 gallons of oil, 20 per cent. leaked out; what per cent. remained in the cask? how many gallons?

4. A milliner bought a lot of bonnets at $6 apiece, but they being damaged, she was obliged to sell them at a loss of 50 per cent.; how much did she receive apiece for them?

5. My agent sells 4 dozen lamps for me at $9 a dozen: if I allow him 25 per cent. on the sale of them, what per cent. of the money which he receives should he pay over to me? how many dollars?

104. 1. What per cent. of a number is $\frac{3}{8}$ of the number?

Solution. Since the whole of a number equals 100 per cent. of itself, $\frac{3}{8}$ of the number equals $\frac{3}{8}$ of 100 per cent., or $37\frac{1}{2}$ per cent of it; therefore, etc.

2. What per cent. of a day is $\frac{2}{5}$ of it? $\frac{1}{3}$ of it?

3. What per cent. of a bushel is $\frac{4}{25}$ of it? $\frac{1}{50}$ of it?

4. A wealthy gentleman gave $\frac{3}{4}$ of his property to his children, and the rest to his church; what per cent. of it did he give to his church?

5. A man's house is insured for $\frac{2}{3}$ of its value; what per cent. of its value does the insurance cover?

6. What is the per cent. of gain when cloth is sold for $\frac{3}{2}$ of what it cost? for $\frac{4}{3}$? for $\frac{6}{5}$? for $\frac{7}{5}$? for $\frac{11}{8}$?

7. What is the per cent. of loss when an article is sold for $\frac{3}{4}$ of its cost? for $\frac{4}{5}$? for $\frac{5}{6}$? for $\frac{5}{8}$?

8. A lawyer received $2 for collecting a bill of $10; what per cent. of the money collected did he receive?

Suggestion. $2 is $\frac{2}{10}$ or $\frac{1}{5}$ of the $10 collected; $\frac{1}{5}$ equals $\frac{1}{5}$ of 100 per cent., or 20 per cent.

9. 12 soldiers of a detachment of 40 soldiers were placed upon guard; what per cent. of the detachment were placed upon guard? what per cent. remained?

10. What per cent. of $40 is $12? is $20? is $15?

11. If I sell boots at $6 a pair which cost me $5 a pair, how many dollars do I gain on a pair? what per cent.?

NOTE. The per cent. of gain or loss should be reckoned on the money which I expended.

12. By selling coal at $8 a ton which cost $10, how many dollars do I lose on a ton? what per cent.?

13. I bought apples at $3 a barrel; if I sell 10 barrels of them for $40, what per cent. do I gain on the 10 barrels?

105. 1. An agent has a commission of 20 per cent. for selling goods; if his commission is $12, what is the amount of his sales?

Suggestion. 20 per cent. is $\frac{20}{100}$, or $\frac{1}{5}$; $12 is $\frac{1}{5}$ of $60.

2. A trader sold a horse for $12½ more than he cost, and thereby gained 10 per cent.; what did he cost?

3. 12½ is 10 per cent. of what number?

4. I have received $4½ for collecting a bill, which is 25 per cent. of the sum collected; what was the sum collected?

5. Mr. Goodwin has 18 cows more this year than he had last year, which is 30 per cent. of what he then had; how many cows had he last year?

6. 18 is 30 per cent. of what number?

7. A man failing in business can pay but 40 per cent. of his debts; if he pays me $15, how much did he owe me?

8. $15 is 40 per cent. of what number?

9. A farmer sold peaches at $3 a bushel, which was

25 per cent. less than the market price; what was the market price?

Suggestion. If \$3 is 25 per cent. less than the market price, it must be 75 per cent., or $\frac{3}{4}$ of the market price.

10. After losing 20 per cent. of his potatoes by disease, a farmer had 100 bushels left; how many bushels had he before he lost any?

11. William sells pineapples for 22 cents each, which is $8\frac{1}{3}$ per cent. less than they cost; what did they cost?

12. A broker sells stock at \$30 a share, which is 40 per cent. less than its original value; what was its original value?

13. What did raisins cost a pound which sell for 15 cents a pound, at a gain of 25 per cent.?

Suggestion. If the gain is 25 per cent. or $\frac{1}{4}$ of the cost, 15 cents, which includes both the cost and the gain, must equal $\frac{5}{4}$ of the cost.

14. If there is a gain of 40 per cent. when paper is sold at 21 cents a roll, what was the original cost?

15. Mr. Vose sold land for \$32 an acre, which was 60 per cent. more than he gave for it; what did he give for it?

16. A rose bush has 28 roses upon it to-day, which is $16\frac{2}{3}$ per cent. more than it had yesterday; how many roses had it yesterday?

17. A merchant lost $12\frac{1}{2}$ per cent. on the cost of a harness by selling it at \$42; what did it cost?

18. For what should the above harness have been sold to gain $12\frac{1}{2}$ per cent.?

19. By selling muslin at 18 cents a yard, I lost 10 per cent.; what should I have sold it for, to gain 10 per cent.?

20. A whip and a bridle together cost \$6; if the cost of the whip was equal to 20 per cent. of the cost of the bridle, what was the cost of each?

Section XLII.

INTEREST.

106. A having had the use of $500 of my money for 1 year, pays me $\frac{10}{100}$ of $500, equal to $50, for the use of it. Money paid for the use of money is **Interest.**

Interest is reckoned at a certain *rate per cent.* of the money used. The money used is the **Principal.**

The principal with the interest is the **Amount.**

1. Find the interest of $100 for 1 year at 6 per cent.

Solution. 6 per cent. is $\frac{6}{100}$; $\frac{1}{100}$ of $100 is $1, and $\frac{6}{100}$ is $6; therefore, etc.

2. Find the interest of $100 for 1 year at 7 per cent.
3. Find the interest of $200 for 1 year at 5 per cent.
4. Find the interest of $500 for 1 year at 12 per cent.
5. Find the interest of $150 for 1 year at 1 per cent; at 6 per cent.; at 4 per cent.; at 5 per cent.
6. Find the interest of $250 for 1 year at 1 per cent.; at 3 per cent.; at 7 per cent.
7. Find the interest of $225 for 1 year at 1 per cent.; at 2 per cent.; at 8 per cent.; at 10 per cent.
8. Find the amount of $400 for 1 year at $7\frac{1}{2}$ per cent.
9. Find the amount of $500 for 1 year at $7\frac{3}{10}$ per cent.
10. Find the interest of $100 at 7 per cent. for 1 year; for 2 years; for 2 y. 6 mos. ($2\frac{1}{2}$ years).
11. Find the interest of $300 at 6 per cent. for 2 y. 3 mo.; for 3 y. 8 mo.
12. Find the interest of $50 at 8 per cent. for 2 y. 4 mo.; for 3 y. 2 mo.
13. Find the interest of $60 at 5 per cent. for 2 y. 9 mo.; for 4 y. 2 mo.
14. Find the amount of $140 at 5 per cent. for 3 y. 10 mo.; for 1 y. 5 mo.
15. Find the interest of $300 for 1 y. 1 mo. 15 d. at 8 per cent.

Solution. The interest of $300 for 1 year is $24; for 1 month, $\frac{1}{12}$ of $24, or $2; and for 15 days, $\frac{1}{2}$ of $2, or $1: $24 plus $2, plus $1, equals $27. *Ans.* $27.

16. Find the interest of $200 for 2 y. 1 mo. 15 d. at 6 per cent.

17. Find the interest of $100 for 4 y. 1 mo. 20 d. at 9 per cent.

18. Find the interest of $600 for 1 y. 2 mo. 6 d. at 3 per cent.

19. Find the interest of $120 for 1 y. 3 mo. 3 d. at 10 per cent.

20. Find the interest of $450 for 1 y. 6 mo. 1 d. at 4 per cent.

21. Find the amount of $50 for 3 y. 2 mo. 24 d. at 12 per cent.

107. TO COMPUTE INTEREST AT 6 PER CENT.

At 1 per cent., the interest of any sum of money will equal the principal in 100 years.

1. If the interest equals the principal in 100 years at 1 per cent., in how many years will it equal the principal at 2 per cent.? at 3 per cent.? at 4 per cent.? at 5 per cent.? at 6 per cent.?

2. If at 6 per cent. the interest equals the principal in 16⅔ years, or 200 months, what part of the principal will the interest equal in 100 months? in 50 months? 25 months? 12½ months?

3. What is the interest of $50 for 200 months? of $60 for 100 mo.? of $80 for 50 mo.? of $72 for 25 mo.? of $32 for 12½ mo.?

4. What part of the principal will the interest equal in 40 months? 20 mo.? 10 mo.? 66⅔ mo.? 33⅓ mo.?

5. What is the interest of $100 for 40 months? of $150 for 20 mo.? of $40 for 10 mo.?

6. What is the interest of $33 for $66\frac{2}{3}$ months? of $60.36 for $33\frac{1}{3}$ mo.? of $144.12 for $16\frac{2}{3}$ mo.?

7. What part of the principal will the interest equal in 2 months?

8. What is the interest of $100 for 2 months? of $250? of $17? of $103? of $1?

9. If the interest equals $\frac{1}{100}$ of the principal in 2 months, how many hundredths of the principal will the interest equal in 4 months? in 6 mo.? 8 mo.? 10 mo.? 12 mo.? 14 mo.?

10. What part of $\frac{1}{100}$ of the principal will the interest equal in 1 month? in 3 mo.? 5 mo.? 7 mo.?

11. What is the interest of $20 for 2 months? of $40 for 3 mo.? of $18 for 4 mo.? of $100 for 8 mo.? of $12 for 9 mo.? of $1 for 1 mo.?

12. If the interest equals $\frac{1}{100}$ of the principal in 60 days (2 months), what part of $\frac{1}{100}$ of the principal will the interest equal in 20 days? in 15 days? 12 days? 10 days? 6 days? 5 days? 3 days? 2 days? 1 day?

13. What is the interest of $1000 for 60 days? 30 days? 15 days? 12 days? 6 days? 3 days? 1 day?

14. What is the interest of $100 for 1 month 3 days?

Solution. The interest of $100 for 1 month is $\frac{1}{2}$ of $\frac{1}{100}$ of $100, or $.50; the interest of $100 for 3 days is $\frac{1}{10}$ of the interest for 30 days, or $\frac{1}{10}$ of $.50, which is $.05; $.50+$.05=$.55. *Ans.* $.55.

15. What is the interest of $200 for 2 months 3 days? for 3 mo. 3 d.? for 4 mo. 3 d.? for 5 mo. 3 d.?

16. What is the interest of $160 for 4 years 2 months 11 days (4 yr. 2 mo. = 50 mo.)?

17. What is the interest of $180.60 for 2 years 9 months 10 days ($33\frac{1}{3}$ months)?

18. What is the amount of $600 for 3 years 4 months 20 days?

19. What is the amount of $330 for 5 yr. 6 mo. 20 d.?

108. The following examples illustrate a method of computing interest at 6 per cent. which is in general use.

If at 6 per cent. the interest for 2 months is $\frac{1}{100}$ of the principal, the interest of $1 for 2 months will be 1 cent; and for 6 days, which is $\frac{1}{10}$ of 2 months, it will be 1 mill. Thus the interest of $1 for 4 months is 2 cents, for 1 month $\frac{1}{2}$ cent, for 3 months $1\frac{1}{2}$ cents, for 5 months $2\frac{1}{2}$ cents, for 6 months 3 cents, etc.; and the interest of $1 for 12 days is 2 mills, for 18 days, 3 mills, for 3 days, $\frac{1}{2}$ of a mill, for 2 days, $\frac{1}{3}$ of a mill, for 7 days, $1\frac{1}{6}$ mills, etc.

1. Find the interest of $1 for 4 months; of $5 for 4 months.

Solution. The interest of $1 for 4 months is 2 cents, and of $5 it is 5 times 2 cents, or 10 cents; therefore, etc.

2. Find the interest of $1 for 1 month; of $10 for 1 mo.; of $6 for 3 mo.; of $12 for 5 mo.

3. Find the interest of $1 for 6 days; of $20 for 6 days; of $30 for 12 days; of $10 for 18 days.

4. Find the interest of $1 for 3 days; of $20 for 3 days; of $30 for 2 days; of $60 for 1 day.

5. Find the interest of $1 for 9 days; of $100 for 9 days; of $200 for 15 days; of $600 for 7 days.

6. Find the interest of $4 for 1 year 3 months 3 days.

Suggestion. Find the interest for years, for months, and for days separately, and add the several items.

7. Find the interest of $10 for 2 y. 2 m. 18 d.

8. Find the interest of $12 for 2 y. 1 m. 15 d.

9. Find the interest of $15 for 1 y. 4 m. 9 d.

10. Find the interest of $20 for 3 y. 6 m. 12 d.

NOTE. To find the interest by the above process at any other rate than 6 per cent., first find the interest at 6 per cent., and then increase or diminish that interest as the given per cent. is greater or less than 6 per cent. Thus, at 7 per cent. take $\frac{7}{6}$ of the interest at 6 per cent., or add $\frac{1}{6}$; at 5 per cent. take $\frac{5}{6}$ of the interest at 6 per cent., or subtract $\frac{1}{6}$, etc.

109. 1. In what time will $100 gain $12 of interest at 6 per cent.?

Solution. The interest of $100 for 1 year at 6 per cent. is $6; it will require as many years for $100 to gain $12 as there are 6's in 12; there are two 6's in 12. *Ans.* 2 years.

What time will be required —

2. For $100 to gain $10 at 5 per cent.?
3. For $200 to gain $18 at 6 per cent.?
4. For $300 to gain $16 at 3 per cent.?
5. For $30 to gain $8 at 4 per cent.?

110. 1. At what rate per cent. will $100 gain $6 in 2 years?

Solution. The interest of $100 for 2 years at 1 per cent. is $2; it will require as many 1 per cent.'s to gain $6, as there are 2's in 6; there are three 2's in 6. *Ans.* 3 per cent.

At what rate per cent. —

2. Will $100 gain $10 in 2 years?
3. Will $400 gain $24 in 1 year?
4. Will $120 gain $30 in 5 years?
5. Will $125 gain $30 in 4 years?

111. 1. What principal will gain 60 cents of interest in 2 years 6 months at 8 per cent.?

Solution. The interest of $1 for 2 years 6 months at 8 per cent. is 20 cents: it will require as many dollars of principal to gain 60 cents of interest, as there are 20's in 60; there are three 20's in 60. *Ans.* $3.

What principal —

2. Will gain $.80 in 4 years at 2 per cent.?
3. Will gain $1 in 3 years 4 months at 6 per cent.?
4. Will gain $.50 in 2 months at 12 per cent.?
5. Will gain $1.44 in 8 months at 9 per cent.?

Section XLIII.

GENERAL REVIEW.

112. 1. A merchant sold 5 barrels of sugar at $8 a barrel, and thereby lost $15; what did the sugar cost him a barrel?

2. John and James found $30: after paying $2 for advertising it, they divided the balance equally between themselves; how much did each boy have?

3. A shoe dealer charged $80 for 4 cases of boots at $2 a pair; how many pairs were there in each case?

4. A cistern that will hold 96 gallons has 2 pipes: by the first it receives 3 gallons in 1 hour, and by the second it receives 5 gallons in 1 hour; in how many hours will the cistern be filled?

5. If two vessels start from the same place at the same time, and sail in opposite directions, one at the rate of 8 and the other at the rate of 5 miles an hour, how far apart will they be at the end of 1 hour? In how many hours will they be 52 miles apart?

6. An accommodation train and an express train start from the same station and run on the same route, one at the rate of 16 miles an hour and the other at the rate of 40 miles an hour; how far apart will they be at the end of 2 hours?

7. A merchant ship sails from port at the rate of 7 miles an hour: 4 hours after, a privateer starts from the same port and follows her at the rate of 10 miles an hour; in what time will the privateer overtake the merchantman?

8. The hind wheel of a carriage is 15 feet in circumference, and the forward wheel is 10 feet in circumference; in going 120 feet, how many more times will the forward wheel turn than the hind wheel?

113. 1. At \$6 a cord, how many cords of wood must be given for 4 dozen rakes at \$9 a dozen?

2. If 3 apples are worth 6 cents, how many apples must be given for 8 pears at 3 cents apiece?

3. If 3 sheep are worth \$12, how many sheep must be given for 8 cwt. of salt at \$1$\frac{1}{2}$ a cwt.?

4. How many dozen forks, at \$6 a dozen, will pay for 8 hats, if 10 hats are worth \$30?

5. If \$11 will buy 77 yards of muslin, how many yards will \$5 buy?

6. A man bought 10 gallons of oil for \$30, and gave 8 gallons of it for 2 ploughs; what was the price of 1 plough?

7. If the interest of \$10 for a certain time is \$1.20, what is the interest of \$12 at the same rate and for the same time?

8. If 8 men can do a piece of work in 5 days, how many men can do the same work in 4 days?

9. If 3 men can dig a trench in 8 days, in how many days can 6 men dig it?

10. If a cistern can be emptied by 2 pipes in 8 hours, in what time can it be emptied by 5 pipes of the same size?

11. If a staff 14 feet high casts a shadow of 7 feet, how high is that staff which casts a shadow of 11 feet at the same time of day?

114. 1. If 5 men eat 20 pounds of meat in 8 days, how many pounds will 6 men eat in 10 days?

Solution. If 5 men eat 20 pounds of meat, 1 man will eat $\frac{1}{5}$ of 20 pounds, or 4 pounds, and 6 men will eat 6 times 4 pounds, or 24 pounds; if the men eat 24 pounds in 8 days, in 1 day they will eat $\frac{1}{8}$ of 24 pounds, or 3 pounds, and in 10 days they will eat 10 times 3 pounds, or 30 pounds; therefore, etc.

2. If it costs $36 to keep 3 horses for 4 weeks, what will it cost to keep 7 horses for 5 weeks?

3. If 7 men can build 70 rods of wall in 5 days, how many rods can 3 men build in 4 days?

4. If the rent of 7 houses for 2 months is $84, what is the rent of 5 of the houses for 3 months?

5. If the interest of a certain sum for 5 years at 4 per cent. is $40, what is the interest of the same sum for 2 years at 10 per cent.?

6. If 2 pipes will empty a cistern containing 4 barrels in 6 hours, in what time will 3 pipes empty a cistern containing 12 barrels?

7. If 12 yoke of oxen can remove a certain quantity of stone in 4 days, how many oxen will be required to remove twice as much stone in 6 days?

8. If 10 men can remove a wall in 4 days, in what time can 8 men rebuild it, if it takes 3 times as long to rebuild it as to remove it?

115. 1. What number is that out of which if you take $\frac{1}{2}$, the remainder will be $\frac{1}{3}$?

2. What number is that to which if you add $1\frac{3}{4}$, the sum will be $3\frac{1}{2}$?

3. From a piece of muslin containing $34\frac{1}{2}$ yards, $10\frac{3}{4}$ yards have been cut; how much more must be cut from it, that $10\frac{2}{3}$ yards may remain?

4. After paying $\$10\frac{3}{5}$ for a barrel of flour, and $\$6\frac{5}{6}$ for a tub of lard, I have just $\$7\frac{9}{10}$ remaining; how much money had I at first?

5. If a piece of work can be done by 5 men in 1 day, how many men must be employed to do the same work in $\frac{1}{2}$ of a day? in $\frac{1}{3}$ of a day?

6. If a man can walk $5\frac{1}{2}$ miles in $\frac{1}{6}$ of a day, how far can he walk in 1 day?

7. If it requires $10\frac{1}{3}$ yards of carpeting which is 1

yard wide to carpet a floor, how much carpeting which is $\frac{1}{2}$ of a yard wide will be required to carpet the same floor?

8. If $12\frac{5}{10}$ feet of gas can pass through a burner in $\frac{1}{5}$ of a minute, how much can pass through the same burner in 1 minute?

9. Charles tied up 8 packages, using $2\frac{1}{3}$ feet of cord for each package; how many feet of cord did he use?

10. What is the cost of $\frac{2}{3}$ of a yard of linen at $\$\frac{5}{6}$ a yard, $1\frac{1}{2}$ yards of silk at $\$1\frac{1}{8}$ a yard, and 1 pair of gloves at $\$1\frac{3}{4}$ a pair?

11. A load of hay, with the wagon, weighed $2\frac{3}{4}$ tons; if $\frac{1}{3}$ of this was the weight of the wagon, what was the weight of the hay?

12. If $\frac{1}{11}$ of a cistern can be emptied in 5 minutes by 2 pipes, how many pipes will be required to empty the rest of the cistern in the same length of time?

13. A grocer had a tub of butter containing 54 pounds, and sold $\frac{2}{9}$ of it for $\$4\frac{4}{5}$; what was the price of the butter a pound?

14. A man had 72 acres of land, and divided $\frac{5}{8}$ of it among his 4 sons; how many acres did each son receive?

15. What sum of money must be divided among 9 persons, that each may receive $\$11\frac{7}{10}$?

16. If 6 acres of land can be ploughed in $\frac{1}{2}$ of a day, in what time can 15 acres be ploughed?

17. If 5 pounds of cheese cost 60 cents, what will be the cost of 4 pounds 2 ounces ($4\frac{1}{8}$ lbs.)? 12 lbs. 12 oz.?

18. If 3 yards of silk cost $\$5\frac{5}{8}$, what will 11 yards cost? $1\frac{1}{3}$ yards?

19. A farmer having $25\frac{3}{5}$ acres of land, rented $\frac{1}{2}$ of it, and sold $\frac{1}{2}$ of the remainder at $100 an acre; what did he receive for the part which he sold?

20. 5 times a certain number equals $62\frac{1}{2}$; what is the number?

21. What is that number which contains $2\frac{1}{3}$, 3 times?

22. What number divided by 6 will give $2\frac{1}{3}$ for the answer?

23. What number divided by $\frac{1}{3}$ will give 4 for the answer?

24. $\frac{2}{3}$ of 9 times the age of William equals 54 years; what is his age?

25. If a man can do $\frac{1}{11}$ of a piece of work in $2\frac{1}{2}$ hours, in what time can he do the whole work?

26. If in $\frac{2}{3}$ of a bushel there are $21\frac{1}{3}$ quarts, how many quarts are there in 1 bushel?

27. Jane can do $\frac{7}{8}$ of a piece of work in a day; if this is $\frac{5}{6}$ of what Wilbur can do, how much can Wilbur do in 1 day?

28. A man sold an ox for $48, and thereby lost $\frac{1}{5}$ of what he paid for him; if he paid for the ox with butter at $8 a firkin, how many firkins did he give?

116. 1. How many cents must I pay for 1 book at the rate of $6 a dozen? at $3 a dozen? at $9? at $$4\frac{1}{2}$? $8? $4? $15? $$7\frac{1}{2}$?

2. What part of 1£ is 15 s.? 5 s.? 2 s. 6 d.?

3. What is the cost of 2 lbs. 10 oz. of meat at 24 cents a pound?

4. Mr. Oaks has 5 lbs. 3 oz. of silver ware, and pays a tax of 5 cents an ounce on all but 4 ounces; what tax does he pay for his silver?

5. How many powders can a druggist make from 1 ounce if he puts 10 grains into each powder?

6. How many rods of fencing will be required to enclose a rectangular garden 33 yards long and 22 yards wide? How many square rods will the garden contain?

7. Two rectangular fields are of the same length: one is $13\frac{1}{2}$ rods wide, and contains $1\frac{1}{2}$ acres; if the other is $11\frac{1}{4}$ rods wide, how many acres does it contain?

8. At 7 cents a quart, what is Mr. Snow's milk bill for 1 week, if he takes 3 pints every day, and 2 pints extra on Saturday?

9. Mr. Stratton's two children picked and sent to market 1 bushel of berries, for which they received 10 cents a quart; what was each child's share of the money?

10. If 12 words are transmitted by the Atlantic Telegraph in 1 minute, how long will it take to transmit a message of 96 words? If the Company receives $5 for each word transmitted, how many dollars will they earn in 1 minute? in 10 minutes?

11. How many square rods are there in an acre? What must be the length of a park which is 10 rods wide, to contain an acre?

12. What must be the length of a piece of board which is 9 inches wide, to contain a square foot?

13. If the surface of a piece of marble contains 10 square feet, and its length is 4 feet, what is its width?

14. How many yards of carpeting 1 yard wide will carpet a floor which is 11 feet long and 10 feet wide?

15. If it takes 9 yards of silk that is 1 yard wide to make a garment, how many yards that is $\frac{7}{8}$ of a yard wide will it take to make the same garment?

16. How many yards of silk $\frac{1}{2}$ of a yard wide will line $7\frac{1}{5}$ yards of broadcloth that is $1\frac{1}{4}$ yards wide?

17. How many cubic feet in a rectangular block whose length is 2 feet, whose breadth is $1\frac{1}{2}$ feet, and whose height is 1 foot? How many yards of carpeting 1 yard wide will be required to cover the top and sides of the block?

18. How many square feet of paper will be required to line the bottom and sides of a box that measures on the inside 5 feet in length, 2 in width, and $1\frac{1}{2}$ in height? How many yards of paper $\frac{1}{2}$ of a yard wide will be required to line the same?

19. What will be the cost of a pile of wood 4 feet wide, 4 feet high, and 20 feet long, at $9 a cord?

20. How many cords of wood in a pile 12 feet long, 4 feet wide, and 6 feet high?

21. If a block of granite which contains 60 cubic feet is 4 feet wide and 2 feet high, what is its length?

22. A certain room contains 60 cubic yards: if its length is 5 yards, and its width 4 yards, what is its height?

117. 1. What is $\frac{2}{3}$ of 6 per cent. of 25 rods?

2. $\frac{2}{5}$ of $\frac{5}{6}$ of 30 per cent. of a certain sum is $50; what is the sum?

3. A man owned $\frac{3}{4}$ of a field of wheat; if he sold $33\frac{1}{3}$ per cent. of his share, what per cent. of the field did he still own?

4. What per cent. of 1 dollar is 1 dime?

5. What per cent. of 1 pound is 1 shilling?

6. The crew of a vessel consisted of 10 men, which was equal to 40 per cent. of the lady passengers and 25 per cent. of the gentlemen passengers; how many passengers were on board the vessel?

7. The difference between 25 per cent. and $16\frac{2}{3}$ per cent. of a man's age is 6 years; what is his age?

8. A steamer which was running 12 miles an hour, increased her speed 20 per cent.; how many miles did she then run in an hour?

9. What must be paid for 1 dozen school-books marked at 50 cents apiece, if 20 per cent. of the price is deducted for cash?

10. A book is marked to be sold at $1.25, and is charged to a retailer at 20 per cent. below this price; if, for paying cash, 5 per cent. of the price at which the book is charged to him is deducted, what will the retailer actually pay for the book?

11. A book is sold for 81 cents: if that price is 10 per cent. less than the price at which it is marked, at what price is it marked?

12. A jeweller gained 30 per cent. by selling a watch for $15 more than it cost him; what did it cost him?

13. By a pipe 9 gallons ran out from a cistern; if this was 15 per cent. of what the cistern contained at first, how many gallons remained in the cistern?

14. $27 is $12\frac{1}{2}$ per cent. more than what sum?

15. 45 men is 10 per cent. less than what number?

16. A grocer sells tea at 48 cents a pound, and thus gains 20 per cent.; what did he give a pound for the tea?

17. A miller sells flour at $8 a barrel which cost $5; what per cent. does he gain?

18. A man sold a house for $1800, which was 10 per cent. less than it cost him; what did it cost him? For how much should he have sold it to gain 25 per cent.?

19. How many wringers at $6 apiece must I buy, that, by selling them at a profit of $33\frac{1}{3}$ per cent. I may gain $10?

20. A hatter sold 2 hats at $6 apiece: on one he realized a profit of 50 per cent., and on the other he lost 25 per cent.; what did he gain or lose on both hats?

21. What is the interest of $25 for 2 years 6 months at 4 per cent.? what is the amount?

22. What is the amount of $80 for 3 months 3 days at 6 per cent.?

23. At 6 per cent., what is the interest of $1000 for 3 days? of $500? of $300?

24. At what per cent. must $100 be on interest 4 months to gain $2?

25. For how many years must $50 be on interest at 7 per cent. to gain $14?

26. What principal must be on interest 6 months at 8 per cent. to gain 10 cents?

Section XLIV.

MISCELLANEOUS EXAMPLES.

118. 1. A can do a piece of work in 2 days, and B can do the same work in 3 days; what part of the work can A do in 1 day? what part of it can B do in 1 day?

2. If A can do $\frac{1}{2}$ of a piece of work in 1 day, and B can do $\frac{1}{3}$ of it, what part of the work can both together do in 1 day? How long will it take both together to do the whole work?

3. Mr. Jones can cut a lot of wood in 2 days, and his son can cut the same lot in 4 days; what part of the lot can both together cut in 1 day? In how many days can both together cut the lot?

4. A cistern has 2 outlets: by one it can be emptied in 3 minutes, and by the other in 5 minutes; in how many minutes can both together empty it?

5. A can reap a field in 3 days, and B can reap it in 4 days; in what time can both together reap it?

6. A quantity of earth can be removed by men in 12 days, by horses in 4 days, and by a steam engine in 2 days; in what time can it be removed by all together?

7. A can dig $\frac{1}{2}$ of a trench in 4 days, and B can dig $\frac{1}{3}$ of it in 2 days; how long would it take for A to dig the whole? for B? How many days would it take both together to dig the whole?

8. If John can eat $\frac{1}{5}$ of a barrel of apples in 2 weeks, and Rollo can eat $\frac{1}{4}$ of a barrel in 5 weeks, in what time can both eat 1 barrel?

9. If 1 pound of tea will last a man and his wife together 2 weeks, and if it will last the man alone 6 weeks, how long would it last the woman alone?

NOTE. The woman drinks $\frac{1}{2}-\frac{1}{6}$, which is $\frac{1}{3}$ of the tea, in 1 week.

10. Mary and Ellen together can do the washing for a certain family in 4 hours: if Ellen can do it alone in 6 hours, in what time can Mary do it alone?

11. Lyman can make a case of boots in $\frac{1}{3}$ of a month, and Walter can do the same work in $\frac{1}{2}$ of a month; how many cases can both make in 1 month? How long would it take both together to make 1 case?

12. A can do a piece of work in $\frac{1}{5}$ of a day, and B can do the same work in $\frac{1}{6}$ of a day; how long would it take both to do it?

119. 1. A pencil and a pen together cost 9 cents: if the pencil costs twice as much as the pen, what does each cost?

Solution. 9 cents equals the cost of the pen plus twice the cost of it, which is 3 times the cost of it; hence, $\frac{1}{3}$ of 9 cents or 3 cents is the cost of the pen, and $\frac{2}{3}$ of 9 cents or 6 cents is the cost of the pencil.

2. A man sold a cane and a knife for 8 dollars: for the cane he received 3 times as much as for the knife; what did he receive for each?

3. What number added to twice itself gives 21?

4. The sum of the ages of a brother and sister is 24 years, and the brother is 3 times as old as the sister; how old is each?

5. A and B receive $20 for mowing a field; if A receives 4 times as much as B, what part of the money does each receive?

6. Divide $24 among A, B, and C, so that B shall have 2 times as much as A, and C shall have as much as both A and B.

7. A man dying, left $50,000, as follows: to each of his two nephews a certain sum, to his son 4 times as much as to each of his nephews, and to his wife $2000 more than to his son and nephews together; how much did he leave to each?

120. 1. There are 26 fowls in a yard: 2 more than one half of them are hens, and the rest turkeys; how many are turkeys?

2. A man sold a chair for $14.75, which was $2.25 more than double what it cost him; what did it cost?

3. A book and a map together, cost $5; if the book cost $1 more than the map, what did each cost?

NOTE. If $1 is taken out of $5, the remainder will equal the cost of two books.

4. Upon a certain tree there are 3 more sparrows than robins, and there are 15 birds of both kinds; how many are there of each?

5. A certain class consisted of an equal number of girls and boys; if, after 4 boys had left, 18 pupils remained, how many of each sex were in the class at first?

6. The sum of two numbers is 12, and one of them is 2 more than the other; what are the numbers?

7. The sum of two numbers is 17, and the difference is 3; what are the numbers?

8. In a school of 80 pupils there are three classes: the first and second classes have an equal number of pupils, and the third class has 5 more than the first or second; how many are there in each class?

9. The profits on a book are 45 cents; how can this sum be divided so that the publisher shall have 4 times as much as the author, and the retailer shall have 5 cents more than the author and publisher together?

121. 1. What number is that to which if $\frac{1}{5}$ of itself is added, the sum is 66? [See Art. 84, Example 23.]

2. Divide the number 10 into two such parts that the less shall be $\frac{1}{4}$ as large as the greater.

3. Divide 16 apples between Albert and Mary, giving to Mary $\frac{1}{3}$ as many as to Albert; how many apples will each have?

4. How shall A and B divide \$30 between themselves so that B shall have $\frac{2}{3}$ as many dollars as A?

5. A steamer and a locomotive start from a landing at the same time and go in opposite directions; if the steamer goes $\frac{3}{5}$ as fast as the locomotive, how many miles will each have gone when they are 48 miles apart?

6. The top of a tree 64 feet high was broken off in a storm: if the part broken off was equal to $\frac{1}{7}$ of the part standing, what was the length of each part?

7. Divide 1 hour (60 minutes) into two such parts that $\frac{3}{7}$ of the greater part shall equal the less; what are the parts?

8. Divide 28 into two such parts that the second part shall contain $\frac{1}{3}$ more than the other part.

Suggestion. 28 contains the first part plus $\frac{4}{3}$ of the first part, or $\frac{7}{3}$ of it.

9. Jane and Mary divided some cambric between themselves, so that Jane had $\frac{1}{5}$ more than Mary; what part of the cambric did each have?

10. Mr. Mason sold a gold pencil and a pen for \$22; if the price of the pencil was 20 per cent. higher than the price of the pen, what was the price of each?

122. 1. Charles bought an equal number of 2-cent stamps and 3-cent stamps, paying for the whole 15 cents; how many of each kind did he buy?

Solution. For 1 of each kind he would pay 5 cents; hence, for 15 cents he could buy as many of each kind as there are 5's in 15; there are three 5's in 15; therefore, etc.

2. A tailor had 36 yards of cloth, which he wished to cut so as to have an equal number of coats and vests, the coats to contain $6\frac{1}{2}$ yards each, and the vests $2\frac{1}{2}$ yards each; how many of each could he have?

3. A grocer mixes two kinds of sugar in equal quan-

tities, one of which cost 6 cents a pound and the other 5 cents a pound; he sells the mixture for $1.10, and thereby makes 11 cents; how many pounds of each kind of sugar did the mixture contain?

4. A trader has apples which he sells, some for $2 a barrel, some for $3, and some for $4 a barrel; if he sells an equal number of each kind, and receives $72 for the lot, how many of each kind does he sell?

5. A druggist receives an order for an equal quantity of each of three kinds of drugs, one kind being worth $\$\frac{3}{4}$, another worth $\$1\frac{1}{4}$, and the other worth $2 a pound: if he receives $50 for the whole, how many pounds of each kind are required to fill the order?

6. A shoe dealer fills an order for $70 worth of boots, one kind worth $4 a pair, another worth $3 a pair: of the latter kind there are twice as many pairs as of the former; how many pairs of each kind are required?

7. A confectioner bought glass jars, some at $\$1\frac{5}{6}$, and some at $\$2\frac{2}{3}$ apiece, of the former kind 4 times as many as of the latter; if he paid $60 for the lot, how many of each kind did he buy?

123. 1. What is the ratio of 2 to 3; that is, 2 is what part of 3?

NOTE. Ratio means relation. In expressing the relation of two numbers, we express what part one number is of another, or how many numbers equal to one number there are in another; thus, the ratio of 2 to 4 is $\frac{1}{2}$; that is, 2 is $\frac{1}{2}$ of 4; the ratio of 4 to 2 is 2; that is, 4 is two 2's.

2. What is the ratio of 3 to 6? of 6 to 3? of 3 to 4?

3. The ages of two boys are in the ratio of 3 to 4: the age of the younger is 12 years; what is the age of the elder?

NOTE. The ratio of 3 to 4 is $\frac{3}{4}$; hence, 12 years is $\frac{3}{4}$ of the age of the elder boy.

4. Two numbers are in the ratio of 5 to 2; if the larger number is 20, what is the smaller number?

5. Two numbers are in the ratio of 3 to 5, and their sum is 16; what are the numbers?

NOTE. The smaller number is $\frac{3}{5}$ of the larger number; the sum, 16, equals the larger number plus $\frac{3}{5}$ of it, which is $\frac{8}{5}$ of it: 16 is $\frac{8}{5}$ of 10, the larger number, etc.

6. How can you divide $18 between two persons, so that their shares shall be in the ratio of 4 to 5?

7. Divide the number 24 into two parts, which shall be in the ratio of 1 to $\frac{1}{2}$.

124. 1. A and B paid $6 for having three horses shod: if 1 of the horses belonged to A, and 2 belonged to B, what should each pay?

2. Two men hired a threshing machine for $12: if one used it 1 day, and the other used it 3 days, how much should each pay?

3. C and D hired a pasture for $25: C pastured 3 horses, and D pastured 2 horses; how much should each pay?

Solution. Both together pastured 5 horses: since C pastured 3 horses, he should pay $\frac{3}{5}$ of $25, which is $15; and, since D pastured 2 horses, he should pay $\frac{2}{5}$ of $25, which is $10; therefore, C should pay $15, and D should pay $10.

4. Two farmers sold some lambs for $18: one sold 1 lamb, and the other sold 5 lambs; how many dollars should each receive?

5. John and Frank sold some berries for 50 cents; if John sold 6 quarts, and Frank sold 4 quarts, how many cents should each receive?

6. Cyrus had 3 cents and Samuel had 5 cents: with this money they bought 24 marbles; how many marbles should each boy have?

7. Winn had $5, and Webb had $7: they loaned their money for one year, and received 84 cents of interest; how many cents should each boy have?

8. A, B, and C did a piece of work together: A worked 2 days, B worked 3 days, and C worked 4 days; if they received $36 for their work, what was the share of each?

9. Two drovers hired a pasture together for $7: the first put in 4 oxen, and the second 30 sheep; what should each pay if 1 ox eats as much as 10 sheep?

10. 6 men and 6 boys do a piece of work, for which they receive $60; if 3 boys do as much work as 2 men, what will each man receive? what will each boy receive?

Suggestion. 6 boys will do as much work as 4 men.

11. Mr. G. and Mr. B. entered into partnership: Mr. G. put in $2 as often as Mr. B. put in $1; if they lost $75, what was each one's share of the loss?

Suggestion. As often as $3 were put in, Mr. G. put in $2, and Mr. B. $1; therefore, Mr. G. should bear $\frac{2}{3}$ of the loss, and Mr. B. $\frac{1}{3}$.

12. Mr. D. and Mr. F. enter into partnership: the money Mr. D. invests is to the money Mr. F. invests in the ratio of 3 to 4; if they gained $56, what was each one's share?

Suggestion. Mr. D. put in 3 parts, Mr. F. put in 4 parts, and both put in 7 parts.

13. Mr. Chapin gave his two boys $2, and told them to divide the money between them in the ratio of their ages; if one of the boys was 9 years old and the other 11, how much did each boy receive?

14. John, Frank, and Theodore invested their money in apples: John's money was to Frank's as 2 to 3, and Theodore had as much as both of them; if they bought 50 apples, how many should each boy have?

15. X and Y enter into partnership: X puts in $5 and his time, which he values at $10, and Y puts in $10; if they gain $20, what is each one's share?

16. C and D hire a pasture for the summer, paying

$20 for the use of it: C pastures 3 cows for 2 months and D pastures 1 cow for 4 months; what part of the $20 should each pay? how many dollars?

Solution. The pasturing of 3 cows for 2 months is the same as the pasturing of 2 times 3, or 6 cows for 1 month; and the pasturing of 1 cow for 4 months is the same as the the pasturing of 4 cows for 1 month; 6 cows plus 4 cows equal 10 cows. C's share of the cost should be $\frac{6}{10}$, or $\frac{3}{5}$ of $20 = $12, and D's share should be $\frac{4}{10}$, or $\frac{2}{5}$ of $20=$8; therefore, etc.

17. V and W agreed to cut a field of wheat for $39: V sent 2 men for 2 days, and W sent 3 men for 3 days; how much should V and W each receive?

18. The type for a book was set for $144: one office furnished 2 compositors for 7 days, and another 5 for 2 days; how much should each office receive?

125. 1. Homer and Robert together have 10 peaches: $\frac{1}{2}$ of Homer's share equals $\frac{1}{3}$ of Robert's; how many has each?

Suggestion. If $\frac{1}{2}$ of Homer's share equals $\frac{1}{3}$ of Robert's, $\frac{2}{2}$ or the whole of Homer's share will equal $\frac{2}{3}$ of Robert's; and since Robert's share will equal $\frac{3}{3}$ of itself, 10 peaches must equal $\frac{5}{3}$ of Robert's share.

2. A pole 15 feet long was divided into 2 such parts that $\frac{1}{2}$ of one part was equal to $\frac{1}{3}$ of the other; what was the length of each part?

3. The cost of a boat and rigging was $65: $\frac{1}{2}$ of the cost of the boat was equal to $\frac{5}{3}$ of the cost of the rigging; what was the cost of each?

4. $\frac{3}{8}$ of A's age equals $\frac{2}{5}$ of B's age, and the sum of their ages is 31 years; what is the age of each?

5. At what time between one and two o'clock will $\frac{3}{8}$ of the time past one, equal $\frac{1}{4}$ of the time before two?

NOTE. The time from one to two o'clock equals 60 minutes.

6. What is the time in the afternoon when $\frac{2}{3}$ of the time past noon is equal to $\frac{1}{2}$ of the time before midnight?

7. At what time between 4 and 5 o'clock will the hour and minute hands of the clock be together?

Solution. The hands are together at 12 o'clock, between 1 and 2, between 2 and 3, 3 and 4, etc.; they are, therefore, together 11 times in 12 hours, and their first time of meeting after 12 o'clock is at $\frac{1}{11}$ of 12 hours after 12 o'clock, or at 1 o'clock $5\frac{5}{11}$ min.; their second time of meeting after 12 o'clock, is at $\frac{2}{11}$ of 12 hours after 12 o'clock, or at 2 o'clock $10\frac{10}{11}$ min.; and their fourth time of meeting will be at $\frac{4}{11}$ of 12 hours after 12 o'clock, or at 4 o'clock $21\frac{9}{11}$ min.

8. At what time between 6 and 7 o'clock will the hour and minute hands of the clock be together? At what time between 10 and 11? when next?

126. 1. If Alice is 1 year old, and her brother George is 12 years old, in how many years will the age of George be just twice the age of Alice?

Solution. When Alice was born, George was 11 years old: in 11 years from that time Alice will be 11 years old and George will be 22, at which age George will be twice as old as Alice; 11 years from the birth of Alice will be 10 years from the present time; therefore, George will be twice as old as Alice in 10 years.

2. A father is 30 years old and his son is 4; in how many years will the age of the son be just $\frac{1}{2}$ of the age of the father?

3. William and Giles can each earn \$1 in a day: William has already worked 8 days, and Giles has worked 2 days; how many more days must each work before William's money will be just double Giles's?

4. A mother is 40 years old and her son is 22; how many years is it since the son was $\frac{1}{2}$ as old as the mother?

127. 1. By selling a piece of calico at 10 cents a yard which should have been sold at 13 cents, I lost 90 cents, how many yards were there in the piece?

2. Thomas wishes to buy a certain number of chickens: to buy one kind at 25 cents apiece would require $1 more than to buy another kind at 20 cents apiece; how many does he wish to buy?

3. John gave 2 pears apiece to his cousins, and had 5 pears left: if he had given them 4 apiece, he would have wanted 3 more pears; to how many cousins did he give the pears?

Solution. It would have required 8 pears more to give them 4 apiece than to give them 2 apiece; if it required 8 pears more to give 2 more pears to each cousin, there must have been as many cousins as there are 2's in 8, which is 4; therefore, etc.

4. A man bought some bouquets at 25 cents apiece, and had 20 cents left: if he had bought the same number at 27 cents each, he would have wanted 10 cents more to pay for them; how many did he buy?

5. By selling a lot of ploughs at $15 each, a merchant will gain $70; by selling them at $18 each, he will gain $91; how many ploughs are there in the lot?

6. A man bought cloth at the rate of $3 for 4 yards, and sold it at the rate of $7 for 6 yards, and gained $15; how many yards were there?

7. A man agreed to pay a laborer $3 a day for every day he worked, and the laborer was to forfeit $1 for every day he was idle; at the expiration of 10 days he received $22; how many days was he idle?

8. A traveller wishing to be at a certain place at 12 o'clock, finds if he starts at a certain time and goes at the rate of 10 miles an hour he will be 2 hours early, but if he goes at the rate of 6 miles an hour he will be 2 hours late, what is the length of his journey?

128. 1. If to $\frac{1}{60}$ of a person's age, $\frac{1}{3}$, $\frac{1}{4}$, $\frac{1}{5}$, and $\frac{1}{6}$ of his age is added, the sum will be 1 year less than his age, how old is he?

2. There were two equal sums of money on interest, one for 2 years at 3 per cent., and the other for 1 year at 4 per cent.: the interest on both sums was $1.20; what was the interest on each? What was the value of each sum?

3. Suppose a cistern has one pipe which will fill it in 4 hours, and another which will empty it in 3 hours; if the cistern is full, and the water is allowed to run freely through both pipes, in what time will the cistern be emptied?

4. A's hat and coat together cost $15; his hat and vest together cost $13; his vest and coat together cost $18; what was the cost of each article?

5. A and B can do a piece of work in 5 days: B and C can do it in 6 days, and A and C can do it in 10 days; in what time can all do it together? In what time can each do it alone?

6. A boy had a certain number of apples: if he had sold them at the rate of 2 for 1 cent, he would have lost 2 cents; by selling them at the rate of 3 for 2 cents, he gained 7 cents; how many apples had he?

7. A man went to market with a certain number of pigs: the first day he sold $\frac{3}{5}$ of them, the next day he sold 6 pigs more than $\frac{1}{2}$ of the remainder, after which he had 4 pigs left; how many had he at first?

8. When the value of gold coin is 50 per cent. higher than paper money, how many whole dollars in gold can I get for a 10-dollar bill, and how much paper currency besides?

WRITTEN ARITHMETIC.

EXPRESSING NUMBERS.

I. 1. All numbers are expressed by using, singly or in combination, the figures 1, 2, 3, 4, 5, 6, 7, 8, 9, 0.

2. To express any number of units, from one to nine, a single figure is used; thus,

One,	Two,	Three,	Four,	Five,	Six,	Seven,	Eight,	Nine.
1,	2,	3,	4,	5,	6,	7,	8,	9.

3. A collection of ten units is called one *ten*, or a unit of the second order; two such collections, two *tens;* three such collections, three *tens*, etc.

To express any number of tens, from one to nine, the same figures are again used, but they are now placed at the left of another figure, as follows:

1 ten is (expressed thus), 10; 1 ten with 1 unit is (expressed thus), 11;
2 tens are " 20; 2 tens with 9 units are " 29;
3 tens are " 30; 4 tens with 5 units are " 45;
etc. etc.

4. A collection of ten tens is called one *hundred*, or a unit of the third order; two such collections, two *hundred*, etc.

To express any number of hundreds, from one to nine, the same figures are again used, but now they are placed at the left of *two* other figures; as follows:

1 hundred is expressed thus, 100; 2 hundred, 200; 5 hundred, 500;

1 hundred with 1 ten and 1 unit is expressed thus, 111;
2 hundreds with no tens and 4 units are " 204;
9 hundreds with 9 tens and 9 units are " 999;
etc. etc.

5. A collection of ten hundreds is called one *thousand*, or a unit of the fourth order; two such collections, two *thousand*, etc.

To express any number of thousands, from one to nine, the same figures are again used, but they are now placed at the left of *three* other figures ; as follows :

1 thousand is expressed thus, ... 1000 ;
1 thousand with 1 hundred, 1 ten and 1 unit, is expres'd 1111 ;
etc. etc.

From the above, we see that numbers, from single units to thousands, are expressed by a *scale of tens*. All numbers are expressed by the same scale ; thus,

10 units (1st order) = 1 ten (2d order).
10 tens = 1 hundred (3d order).
10 hundreds = 1 thousand (4th order).
10 thousands = 1 ten thousand (etc).
10 ten thousands = 1 hundred thousand.
10 hundred thousands = 1 million.
etc. etc.

From the above, we see that a unit of the second order equals 10 units of the first order ; that a unit of the third order equals 10 units of the second order ; and generally, that a unit of any order equals 10 units of the next lower order.

When whole numbers are expressed by figures, the units' figure occupies the first place at the right, the tens, the second place, the hundreds, the third place, the thousands, the fourth place, etc. ; thus, —

7th. Millions.	6th. Hundred thousands.	5th. Ten thousands.	4th. Thousands.	3d. Hundreds.	2d. Tens.	1st. Units.	Place and order.
8	9	7	8	2	3	2	Number.

232 is read two hundred thirty-two.
8,230 is read eight thousand, two hundred thirty.
78,032 is read seventy-eight thousand, thirty-two.
970,200 is read nine hundred seventy thousand, two hundred.
8,970,000 is read eight million, nine hundred seventy thousand.

2. 1. In what place should a figure be written to represent units? to represent tens? hundreds?

2. Write 2 so that it shall represent tens; so that it shall represent hundreds.

3. How many units are there in 1 ten? in 3 tens? in 1 hundred?

4. How many tens in 20 units? in 60 units? How many tens and units in 15? in 25? in 42?

5. Express in figures, six; forty-six; thirty.

Read the following numbers: —

6. 32.	9. 204.	12. 530.	15. 910.
7. 58.	10. 516.	13. 800.	16. 706.
8. 81.	11. 768.	14. 302.	17. 428.

Express in figures, —

18. Three hundred.
19. Four hundred twenty-eight.
20. Two hundred thirty.
21. Six hundred seven.
22. Eight hundred five.
23. Nine hundred.
24. Seven hundred sixteen.
25. Where should a figure be written to represent thousands?

Read the following numbers: —

26. 3,000.	28. 4,320.	30. 8,901.	32. 6,004.
27. 5,188.	29. 2,032.	31. 7,200.	33. 9,328.

Express in figures, —

34. Four thousand.
35. Four thousand, one hundred.
36. Three thousand, four hundred seven.
37. Five thousand, two hundred sixteen.
38. Seven thousand, thirty-six.
39. Eight thousand, two.
40. Where should a figure be written to represent ten thousands?

Read the following numbers: —

41. 1,807.	43. 10,001.	45. 70,102.	47. 67,300.
42. 3,691.	44. 32,651.	46. 19,013.	48. 40,070.

Express in figures, —

49. One hundred six.
50. Seventeen thousand, eight hundred.
51. Three thousand, forty-five.
52. Seventy thousand, seventy.
53. Four thousand, four.
54. Where should a figure be written to represent hundred thousands?

Read the following numbers: —

55. 183.	57. 36,075.	59. 365,000.	61. 932,001.
56. 1,702.	58. 82,101.	60. 800,000.	62. 700,110.

Express in figures, —

63. One hundred sixty thousand.
64. Fifty-five thousand, seven hundred sixty-two.
65. One hundred six thousand, forty.
66. Three hundred seventy-five thousand, eight hundred.
67. Four hundred thousand, four.
68. Where should a figure be written to represent millions?

Read the following numbers: —

69. 45,843.	72. 1,792,187.	75. 8,673,210.
70. 867,321.	73. 2,223,816.	76. 5,930,021.
71. 3,640,261.	74. 1,001,001.	77. 7,010,102.

Express in figures, —

78. Eighteen thousand.
79. Eight million, one hundred eighty-five.
80. One hundred eighty-five thousand.
81. Two million, two hundred thousand.
82. Fifty-eight thousand, six.
83. Write 5 so that it shall represent tens; so that it shall represent hundreds; thousands; ten thousands.
84. Write 17 so that it shall represent tens; hundreds. *Ans.* 17 tens = 170; 17 hundreds = 1700.
85. Write 45 so that it shall represent tens; hundreds; thousands.

The number of units of the different orders of a number are sometimes called its *terms*.

Thus, in the number 235, the *terms* are 2 hundreds, 3 tens, and 5 units.

86. Name the terms of the number 768.

ADDITION.

3. **Addition** is the process of counting together numbers of the same kind.

EXERCISES IN THE ADDITION OF NUMBERS WHOSE SUM IS NOT GREATER THAN 20.

4. The pupil may copy upon the slate or black-board the figures in each column, and add the numbers as expressed upward from the bottom.

	1.	2.	3.	4.	5.	6.	7.	8.	9.	10.
	2	1	2	4	3	2	2	6	2	5
	5	5	4	3	1	3	3	3	4	1
	4	3	5	6	5	4	5	3	1	6
	3	5	2	1	4	2	3	1	5	2
Sum,	14	14	13							

NOTE. In adding orally, the pupil should mention only the results; thus in example 1, he should not say, 3 and 4 are 7, and 5 are 12, and 2 are 14; but simply, 3, 7, 12, 14.

Proof. We may prove the correctness of the work by adding downward. If the work is right, the same sum will be obtained as before.

NOTE. Every example should be proved, until the pupil is sure he makes no mistakes.

11.	12.	13.	14.	15.	16.	17.	18.	19.
6	4	0	4	1	6	2	0	1
1	1	5	3	5	0	1	3	2
5	2	2	2	6	3	4	4	1
1	1	9	5	1	4	5	7	8
1	7	1	1	3	4	4	3	3

20. Add together 1 and 4 and 4 and 4 and 2 and 2.
21. Add together 2 and 1 and 2 and 2 and 4 and 4.
22. Add together 5 and 3 and 4 and 1 and 3 and 1.
23. Add together 4 and 4 and 1 and 3 and 3 and 2.
24. Add together 1 and 5 and 3 and 4 and 2 and 1.

25.	26.	27.	28.	29.	30.	31.	32.	33.
2	4	1	2	1	5	4	2	2
2	4	3	3	2	1	3	4	3
4	2	1	3	4	3	1	2	4
4	2	4	1	3	2	2	3	5
4	1	3	4	5	3	4	1	2
1	2	5	4	1	3	3	4	1

34.	35.	36.	37.	38.	39.	40.	41.	42.
2	4	2	1	2	0	5	0	1
0	1	5	1	0	2	2	5	3
4	1	1	6	1	2	1	1	2
1	2	3	2	4	2	0	1	3
5	3	4	1	2	5	3	5	2
2	2	0	0	4	7	3	7	6
0	6	1	4	3	0	2	0	0
5	1	4	5	3	2	2	1	3

5. SINGLE COLUMNS, NO SUM EXCEEDING 20.

43.	44.	45.	46.	47.	48.	49.	50.	51.
2	6	6	8	4	5	7	6	2
3	3	9	4	9	0	8	7	7
9	3	8	7	8	4	6	3	6
7	0	7	5	5	3	2	9	9
5	9	2	2	6	8	4	8	8
8	7	6	0	2	6	3	4	6
6	6	3	9	5	6	9	7	7
8	2	8	5	7	9	8	6	8
9	3	0	8	3	0	7	2	7
5	4	2	7	9	3	5	3	8

52. How many are $8 + 6 + 7 + 9 + 8 + 5 + 9 + 7$?
53. How many are $9 + 7 + 8 + 5 + 9 + 6 + 7 + 8$?
54. How many are $5 + 8 + 9 + 6 + 7 + 7 + 8 + 9$?

For further Exercises in the addition of simple units, see Review Circle, p. 176. See, also, Walton's "Dictation Exercises."

6. To Add two or more Orders of Units when the Amount of each Order is less than 10.

Illustrative Example I. Three men were fishing together: one man caught 22 fishes, another 34 fishes, and the third 33 fishes; how many fishes did all catch?

Operation.

Tens.	Units.
2	2
3	4
3	3
Ans. 8	9 fishes.

Explanation. They caught the sum of 22 fishes, and 34 fishes, and 33 fishes. To find the sum, we express the numbers, so that units of the same order shall be expressed in the same column, and draw a line beneath. We then add the units, and find their sum to be 9; we write a figure 9 under the line, for units, in units' place, then add the tens, and find their sum to be 8: we write a figure 8 under the line, in tens' place, and have 89 for the sum; therefore they caught 89 fishes.

Add the following: —

1. Sheep.	2. Cows.	3. Hens.	4. Geese.	5. Eggs.	6. Doves.	7. Ducks.
32	22	18	37	12	231	113
21	13	10	31	30	211	140
16	50	41	20	25	144	433
10	11	30	11	21	312	313

8. How many yards are there in three pieces of cloth, the first containing 32 yards, the second 30, and the third 35 yards?

9. A farmer paid $55 for his horse, $32 for his cow, and $12 for a plough; what did he pay for all?

10. How many pigeons are there in three flocks, if the first contains 212 pigeons, the second 322, and the third 134?

11. A clerk received $26 on Monday, $302 on Tuesday, $221 on Wednesday, $130 on Thursday, $106 on Friday, and $200 on Saturday; how many dollars did he receive during the week?

12. If the smaller of two numbers is 34,527, and the larger 42,371, what is their sum?

7. TO ADD WHEN THE AMOUNT OF ANY ORDER OF UNITS IS 10, OR MORE THAN 10.

Illustrative Example II. A bookseller sold 23 books in the morning, 32 books in the afternoon, and 17 in the evening; how many books did he sell in all?

Operation.

```
      2 3
      3 2
      1 7
Ans.  7 2 books.
```

Explanation. The bookseller sold 23 books, plus 32 books, plus 17 books. We express the given numbers, units under units, tens under tens, and draw a line beneath. Adding the units, we obtain 12 units, which are equal to 1 ten and 2 units. We write a figure 2 under the column of units, and add the 1 ten with the column of tens, making 7 tens. We write a figure 7 under the column of tens, and have 72 for the sum. Therefore, the bookseller sold 72 books.

Add the following: —

1.	2.	3.	4.	5.	6.	7.
Mills.	Men.	Hours.	Sheep.	Boxes.	Boys.	Bees.
39	42	24	28	56	247	783
28	18	61	36	28	674	241
22	31	19	57	9	495	136
17	23	17	82	37	186	235

	8.		9.		10.		11.		12.		13.
14.	392	+	709	+	270	+	7789	+	2187	+	1505
15.	507	+	560	+	369	+	6680	+	7420	+	8858
16.	969	+	615	+	785	+	8932	+	8717	+	6925
17.	368	+	426	+	421	+	3930	+	9161	+	7407

18. In an orchard there are 35 peach-trees and 28 pear-trees; how many trees are there in all?

19. Charles has $45 in the Savings Bank, $38 lent upon interest, and $24 in his purse; how much money has he?

20. How many days are there in the year, there being 92 days in the Spring, 92 in the Summer 91 in the Autumn, and 90 in the Winter?

21. The skull of the human head has 8 bones, the face 14, the ears 4, the tongue 1, and there are 32 teeth; how many bones are there in the head?

22. There are in the trunk 24 spinal bones, 24 ribs, and 5 other bones; how many bones are there in the trunk?

23. In each of the upper extremities there are 32 bones, and in each of the lower extremities there are 30; how many bones are there in all the extremities?

24. How many bones are there in the entire body?

25. A clerk, who spent $597 of his salary during the year, found he had saved $403; what was his salary?

26. A farmer raised 650 bushels of corn in one year, 327 bushels the next year, and 1,218 bushels the third year; how much corn did he raise in the three years?

27. Mr. Austin sold 316 barrels of apples in one year, 235 the next year, 472 the third year, 22 the fourth, and 187 the fifth year; how many barrels did he sell in five years?

28. Abraham Lincoln was born in 1809, and lived 56 years; in what year did he die?

29. Mr. Brown paid $2,500 for a farm, $250 for a horse and carriage, $925 for stock and farming utensils, and $332 for house furnishings; what did he pay for all?

30. Three men entered into partnership: the first man put in $550, the second $2,125, and the third as much as the first and second; how much did all put in?

31. In 1860 there were published in the United States of political papers 372 dailies, 74 bi-weeklies, 84 tri-weeklies, 2,694 weeklies, and 18 others; how many were published in all?

32. In the same year there were published 3,242 political papers, 277 religious, 298 literary, and 234 miscellaneous papers; how many papers were published in all?

33. In the same year there were in the New England States 3,669 miles of railroad; in the Middle Atlantic States 6,321 miles; in the South Atlantic 5,454 miles; in the Gulf States 2,256 miles; in the Interior Southern States 1,806; in the Interior Northern 11,212; and in the Pacific States 73. How many miles of railroad were there in the United States?

SUBTRACTION.

8. Subtraction is the process of taking part of a number away to find how many are left.

The **Subtrahend** is the part taken away.

The **Minuend** is the number, part of which is taken away.

The **Remainder** is the part left.

Illustrative Example I. Jane had 7 apples, and gave away 5; how many had she left? *Ans.* 2 apples. Here, 5 apples are taken out of 7 apples, to find the number Jane had left; 5 is the *subtrahend*, 7 is the *minuend*, and 2 the *remainder*.

9. TO SUBTRACT WHEN EACH TERM IN THE SUBTRAHEND IS NOT GREATER THAN THE CORRESPONDING TERM IN THE MINUEND.

Ill. Ex. II. Charles had 89 cents, and spent 42; how many cents had he left?

Operation.

	Tens.	Units.
Minuend,	8	9
Subtrahend,	4	2
Remainder,	4	7

Explanation. He had left the difference between 42 cents and 89 cents, which is found by taking 42 out of 89. We express the subtrahend under the minuend, units under units, tens under tens, and draw a line beneath. 2 units taken out of 9 units leaves 7 units; we write a 7 under the line, in units' place; 4 tens taken out of 8 tens leaves 4 tens; we write a figure 4 under the line, in tens' place, and have 47 for the remainder; therefore, Charles had 47 cents left.

Proof. Add the remainder to the subtrahend; if the work is correct, the sum will equal the minuend.

	1. Men.	2. Doves.	3. Cows.	4. Cents.	5. Pens.	6. Bees.	7. Oxen.
Minuend,	82	65	96	48	37	74	59
Subtrahend,	31	23	72	15	25	41	8
Remainder,	51						

	8. Rods.		9. Rods.		10. Feet.		11. Feet.		12. Yards.		13. Yards.
14.	39	—	28	16.	267	—	23	18.	8395	—	7232
15.	25	—	15	17.	155	—	13	19.	6184	—	121

20. Out of 5,829 years take 3,708 years.

21. A man who owed $275, has paid $130 of his debt; how many dollars does he still owe?

22. In the year 1850, there were made in the United States 34,253,436 pounds of maple sugar, and 38,863,568 pounds were made in 1860; how many more pounds were made in 1860 than in 1850?

10. To Subtract when any Term in the Subtrahend is Larger than the corresponding Term in the Minuend.

Ill. Ex. III. John had 42 grapes, and gave Mary 27; how many grapes had he left?

Operation.	Tens.	Units.
	(3)	(12)
	4	2
	2	7
Ans.	1	5

In this example, we cannot take 7 units out of 2 units: we will therefore change one of the 4 tens to units, making 10 units: 10 units and 2 units are 12 units; 7 units out of 12 units leaves 5 units. As one of the 4 tens has been taken, but 3 tens are left: 2 tens out of 3 tens leaves 1 ten, and we have 15 for the remainder; therefore John had 15 grapes left.

	1. Nuts.	2. Hogs.	3. Sheep.	4. Men.	5. Boys.	6. Days.	7. Years.
Min.	82	67	31	85	94	50	80
Sub.	56	49	19	38	27	13	67

	8. Rods.		9. Rods.		10. Trees.		11. Trees.		12. Tons.		13. Tons.
14.	248	—	197	16.	362	—	234	18.	4030	—	1632
15.	157	—	49	17.	129	—	86	19.	801	—	584

20. Doves.	21. Doves.	22. Pounds.	23. Pounds.	24. Tons.	25. Tons.
26. 324 —	247	28. 3274 —	1264	30. 92322 —	6283
27. 136 —	178	29. 1732 —	1825	31. 77035 —	1084

11. *Ill. Ex. IV.* Out of 300 take 135.

Operation.

```
     (2) (9) (10)
      3   0   0
      1   3   5
Ans.  1   6   5
```

In this example, as there are no units and no tens in the minuend, we change one of the hundreds to tens, leaving 2 hundreds; 1 hundred equals 10 tens: we change one of the tens to units, leaving 9 tens; 1 ten equals 10 units: 5 units out of 10 units leaves 5 units; 3 tens out of 9 tens leaves 6 tens; 1 hundred out of 2 hundreds leaves 1 hundred, and the answer is 165.

32. Out of 500 take 42; out of 8,005 take 97.

33. Out of 1,000 take 1; out of 10,201 take 536.

34. Claude has 126 cents; how many more cents must he get before he has 300 cents?

35. Mr. Mann had 808 tomato plants, and sold 175 of them; how many had he left?

36. What is the difference between 25,000 and 24,282?

37. A man has 1,805 miles to travel: 1,650 miles are by railroad, and the remainder by stage; how far has he to travel by stage?

38. The Missouri River is 3,096 miles long, and all but 1,216 miles is navigable for steamboats; how far is it navigable for steamboats?

39. The Mississippi River is 3,160 miles long, and is navigable for steamboats 2,070 miles, and for ships 800 miles; how much farther is it navigable for steamboats than for ships? How far is it from the head of ship navigation to the source of the river?

40. Mount Ararat is 12,210 feet high, and Mount Sinai wants 4,713 feet of being as high: what is the height of Mount Sinai?

41. Independence was declared in 1776; how many years have passed since then?

42. The distance from New York to San Francisco by

Cape Horn is 18,000 miles, by the Panama route it is 6,850 miles; how much farther would a person travel in going from New York to San Francisco by Cape Horn than by the Panama route?

31. Take 99 out of 1,000; then take 99 out of what remains, and so on as many times as you can, and tell what finally remains.

ADDITION AND SUBTRACTION COMBINED.

12. 1. Mr. Rowe owes to one man $29, to another $37, to a third $108, and to a fourth $17. He has $47 in bank notes, $50 in silver money, and $155 deposited in the bank; how much money will he have left after his debts are paid?

2. If the smaller of two numbers is 921, and the difference 85, what is the larger number?

3. If the smaller of two numbers is 3,582, and the larger 10,216, what is the difference?

4. If the difference of two numbers is 192, and the larger number 3,201, what is the smaller number?

5. At a threshing match at the World's Exposition in Paris, in 1860, the American Threshing Machine threshed 782 quarts of wheat in 1 half hour, which was 348 quarts more than the English machine threshed in the same time; how many quarts did the English machine thresh in 1 half hour?

6. In 1860, Massachusetts had 157 inhabitants to a square mile, New York 84, Ohio 58, and California 2. How many more inhabitants to a square mile were there in Massachusetts than in New York? than in Ohio? than in California?

7. There are 5,280 feet in a mile: the height of Mount Vesuvius lacks 1,332 feet of being 1 mile; what is the height of Mount Vesuvius?

8. The height of Mount Etna is 314 feet more than 2 miles; what is the height of Mount Etna?

9. Sun-dials were invented 558 years before Christ, and watches 1,477 years after Christ; what time elapsed between these inventions?

10. Wood-cuts were invented 1,460 years after Christ, and daguerreotypes 1,839 years after Christ; what time elapsed between these inventions?

MULTIPLICATION.

13. Multiplication is the process of uniting a number of equal numbers at once.

The **Multiplicand** is one of the equal numbers to be united.

The **Multiplier** is the number which shows how many equal numbers are to be united.

The **Product** is the answer.

Ill. Ex. I. If 1 peach cost 4 cents, what will 3 peaches cost? Here 4 cents, the price of 1 peach, is multiplied by 3, the number of peaches, to obtain the price of all, which is 12 cents. 4 is the *multiplicand*, 3 the *multiplier*, and 12 the *product*.

14. TO MULTIPLY WHEN THE MULTIPLIER CONSISTS OF UNITS ONLY.

Ill. Ex. II. If in 1 day there are 24 hours, how many hours are there in 2 days?

Operation.	
Multiplicand,	24
Multiplier,	2
Product,	48

Explanation. If in 1 day there are 24 hours, in 2 days there are two 24's of hours. We express the multiplier under the multiplicand, and draw a line beneath. We first multiply the units: two 4's of units are 8 units; we write a figure 8 in the units' place, under the line. We next multiply the tens: two 2's of tens are 4 tens; we write 4 in the tens' place, and have 48 for the product.

	1. Apples.	2. Pears.	3. Peaches.	4. Plums.	5. Cherries.	6. Grapes.
Multiply	32	44	323	332	112	221
by	2	2	3	2	4	4
Product,	64					

Ill. Ex. III. What cost 4 acres of land at $36 an acre?

Operation.

Multiplicand,	3 6
Multiplier,	4
Product,	1 4 4

Explanation. If 1 acre costs $36, 4 acres will cost four 36's of dollars; four 6's of units are 24 units, equal to 2 tens and 4 units; we write 4 in the units' place, under the line, and reserve the 2 tens for the tens' place. Four 3's of tens are 12 tens, which, with the 2 tens reserved are 14 tens, equal to 1 hundred and 4 tens; we write 4 in the tens' place and 1 in the hundreds' place, and have 144 for the product.

Proof. Take the multiplicand for the multiplier, and the multiplier for the multiplicand, and multiply again; if the work is correct the product will be the same as before.

	7. Hats.	8. Caps.	9. Gloves.	10. Socks.	11. Coats.	12. Canes.
Multiply	326	908	535	327	548	732
by	3	6	4	9	5	8
Ans.	978	5,448	2,140			

	13. Nails.	14. Saws.	15. Sticks.	16. Screws.	17. Tacks.	18. Rods.
Multiply	492	206	369	835	683	417
by	7	2	6	7	8	9

19. What will 4 brooms cost at 56 cents apiece?

20. What will 5 horses cost at $235 apiece?

21. What will 75 pounds of sugar cost at 8 cents a pound?

NOTE. In multiplying two numbers together, the result is the same whichever number is taken as the multiplier.

The following solution may be given if preferred: —

Solution. 75 pounds of sugar at 1 cent a pound will cost 75 cents; at 8 cents a pound, it will cost eight 75's of cents, etc.

22. What will 107 pounds of pork cost at 9 cents a pound?

23. At $1 a day, what will a person earn in 52 weeks, there being 6 working days in 1 week?

24. At $7 a barrel for flour, what will 1,275 barrels of flour cost?

25. How much must I pay for 15 cords of wood at $6 a cord, and 9 tons of coal at $13 a ton?
26. Multiply 293 by 2; by 3; by 4.
27. Multiply 314 by 3; by 4; by 5.
28. Multiply 425 by 4; by 5; by 6.
29. Multiply 536 by 5; by 6; by 7.
30. Multiply 647 by 6; by 7; by 8.
31. Multiply 758 by 7; by 8; by 9.
32. Multiply 869 by 8; by 9; by 2.
33. Multiply 972 by 9; by 2; by 3.

15. To Multiply by Units of any Order.

Ill. Ex. IV. How many pecks in 10 bushels, each containing 4 pecks? in 100 bushels?

In 10 bu. there are ten 4's of pks. = 40 pks.
In 100 bu. there are 1 hundred 4's of pks. = 400 pks.

From the above we see that *when the multiplier is* 10, *the product may be expressed by annexing to the expression for the multiplicand one zero*; *when the multiplier is* 100, *by annexing two zeros; and so on.*

1. How many yards in 10 pieces of cloth, each containing 33 yards? *Ans.* 330 yards.
2. At $22 a ton, what cost 10 tons of hay? 100 tons?

3. Multiply 236 by 10.	5. Multiply 72 by 1,000.
4. Multiply 706 by 100.	6. Multiply 99 by 10,000.

Ill. Ex. V. If there are 32 quarts in 1 bushel, how many quarts are there in 40 bushels? in 400 bushels?

Operation.

```
       32
      400
Ans. 12,800
```

Explanation. To multiply by 40, which is four 10's, we can multiply by 4, and annex a zero to the expression of the product; to multiply by 400, we can multiply by 4, and annex two zeros to the expression of the product.

7. Multiply 62 by 20.	10. Multiply 1,009 by 700.
8. Multiply 1,804 by 30.	11. Multiply 84 by 9,000.
9. Multiply 989 by 500.	12. Multiply 505 by 60,000.

16. To Multiply when the Multiplier consists of Units, Tens, etc.

Ill. Ex. V. Multiply 38 by 44.

Operation.

```
      3 8
      4 4
    -----
    1 5 2
  1 5 2 tens.
  -------
Ans. 1,6 7 2
```

In this example we first multiply by 4 and obtain, for a partial product, 152 units. We then multiply by 4 tens, and obtain 152 tens for another partial product; this we express as we multiply, so that the first figure shall be in the tens' place; adding the partial products, we obtain for the answer 1,672.

Perform the following: —

	1.		2.			3.		4.			5.		6.
7.	219	×	68.		9.	208	×	126.		11.	927	×	119.
8.	33	×	22.		10.	36	×	57.		12.	64	×	23.

	13.		14.			15.		16.			17.		18.
19.	725	×	372.		21.	149	×	265.		23.	269	×	238.
20.	144	×	25.		22.	62	×	38.		24.	56	×	27.

25. How much wheat will 55 acres yield, if each acre yields 34 bushels?

26. At 138 cents a barrel, what is the cost of freighting 25 barrels of flour?

27. If a house has 27 windows in it, and 12 panes of glass in each window, how many panes of glass are there in all the windows?

28. John sold 25 pigeons at 45 cents apiece; what did he receive for the whole?

Ill. Ex. VII. Multiply 321 by 106.

Operation.

```
   3 2 1
   1 0 6
 -------
  1 9 2 6
 3 2 1
 ---------
 3 4,0 2 6
```

When zeros occur between the other figures of the multiplier, multiply as if there were no zeros, and put the first figure of the expression of each partial product under that of the term by which you multiply.

29. Multiply 1,216 by 207; 825 by 4,002.
30. Multiply 87,000 by 104; 2,005 by 108.

17. EXAMPLES IN ADDITION, SUBTRACTION, AND MULTIPLICATION.

1. At $10 a barrel for flour, what cost 273 barrels, including $1 a barrel for freight?

2. If a boy attends school 6 hours a day for 4 days of each week, and 3 hours a day for 2 days, how many hours will he attend school in 1 week? in 11 weeks?

3. How many visitors can be accommodated in a schoolroom on exhibition day, if there are 14 settees which hold 6 visitors each, 12 which hold 8 visitors each, and 18 chairs?

4. A man bought 12 tubs of butter, each containing 56 pounds, at 33 cents a pound, and 10 cheeses, each weighing 85 pounds, at 13 cents a pound; what should he pay for what he bought?

5. A man sold 35 harnesses at $17 apiece, and 13 wagons, receiving for each wagon 8 times as much as for a harness; what did he receive for all he sold?

6. 7 months of the year have 31 days each, 4 months have 30 days each, and 1 month has 28 days; how many days are there in the whole year?

7. A man bought a horse for $48, and paid $3 a week for keeping him; after 14 weeks, he sold him for $110; what did he gain?

8. One man travels 67 miles in a day, another travels 59 miles; if they start from the same place and travel in opposite directions, how far apart will they be at the end of 1 day? at the end of 11 days? at the end of 16 days?

DIVISION.

18. Division is the process of finding how many equal numbers, one of which is given, there are in another number.

The **Dividend** is the number divided.

The **Divisor** is one of the equal numbers there are in another number.

The **Quotient** is the answer.

Illustrative Example I. How many marbles, at 2 cents apiece can be bought for 13 cents? As many marbles at 2 cents apiece can be bought for 13 cents as there are 2's in 13, which is 6, and 1 remains. Here 13 is the *dividend*, 2 is the *divisor*, 6 is the *quotient*, and 1 is the *remainder*.

19. SHORT DIVISION.

Ill. Ex. II. At 2 cents apiece, how many apples can be bought for 64 cents?

Operation.

```
2) 6 4
   ———
   3 2
```

Explanation. As many apples at 2 cents apiece can be bought for 64 cents as there are 2's in 64. We express in figures the 64 with 2 at the left, drawing a curved line between the expressions, and a straight line under that for the dividend. 64 equals 6 tens, or 60 with 4. In 60 there are thirty 2's, or 3 tens of 2's; we write a figure 3 under the line in the tens' place. In 4 there are two 2's; we write a figure 2 under the line in the unit's place and have 3 tens with 2 units, or 32, for the result; therefore 32 apples can be bought.

Perform the following: —

	1.		2.	3.	4.
Divisor,	2)844	Dividend.	3)369	4)484	5)505
Quotient,	422		123		

Ill. Ex. III. Divide 208 by 4.

Operation.

```
4 ) 208
    ———
     52
```

In the number 2 hundred there are no hundreds of 4's; we therefore first divide 20 tens. In 20 tens there are five 10's of 4's, etc.

5. At $5 a yard, how many yards of broadcloth can be bought for $355?

6. A man has 246 quarts of milk, and wishes to put it into 6-quart cans; how many cans must he use?

7. If a person saves $4 in a week, in how many weeks can he save $328?

Ill. Ex. IV. Divide 615 by 3.

Operation.

```
3 ) 6 1 5
    -----
    2 0 5
```

In this example, after dividing the 6 (hundreds) by 3, we have 1 ten to divide; as in 1 ten there are no tens of 3's, we put 0 in the tens' place of the quotient, and divide 15 units by 3.

8. If 8 gills make 1 quart, how many quarts are there in 832 gills?

Ill. Ex. V. In 1 week there are 7 days; how many weeks are there in 365 days?

Operation.

```
7 ) 3 6 5   Rem.
    -----
      5 2 . 1
```

Here 36 (tens) divided by 7, leaves a remainder of 1 (ten), which, with the 5 units, equals 15 units; again, 15 units divided by 7, leaves a remainder of 1; therefore, in 365 days there are 52 weeks and 1 day remaining.

Proof. Multiply the quotient by the divisor, and to the product add the remainder; the sum should equal the dividend.

9. Divide 8,487 by 2; by 3; by 4.
10. Divide 9,598 by 3; by 4; by 5.
11. Divide 7,609 by 4; by 5; by 6.
12. Divide 1,710 by 5; by 6; by 7.
13. Divide 2,821 by 6; by 7; by 8.
14. Divide 3,932 by 7; by 8; by 9.
15. Divide 4,043 by 8; by 9; by 2.
16. Divide 5,154 by 9; by 2; by 3.
17. How many pounds of sugar at 8 cents a pound can be bought for 196 cents? for 248 cents?
18. How many tons of coal at $9 a ton can be bought for $828? for $593?

20. It is sometimes necessary to find one of the equal parts of a given number; this process is also called Division, and is performed in the same manner as that taught above; thus, to find one half of a number, we divide by 2; to find one third of a number, we divide by 3; to find one fourth, we divide by 4; to find one fifth, we divide by 5; to find one tenth, we divide by 10, etc.

1. What is one half of 18? of 22? of 24?

Ill. Ex. VI. What is one half of 64?

Operation.

$$2\,)\,\underline{6\ 4}$$
$$3\ 2$$

64 equals 6 tens and 4 units; one half of 6 tens is 3 tens, and one half of 4 units is 2 units; therefore, one half of 64 is 32.

2. What is one half of 34? of 82? of 106?
3. What is one third of 18? of 183? of 981?
4. What is one fourth of 64? of 144? of 232?
5. What is one fifth of 55? of 505? of 1,035?
6. What is one sixth of 72? of 96? of 6,738?
7. What is one seventh of 28? of 91? of 7,546?
8. What is one eighth of 88? of 864? of 2,952?
9. What is one ninth of 63? of 234? of 7,263?
10. If 3 coats cost $135, what will 1 coat cost?
11. If 5 shares of railroad stock cost $505, what will 1 share cost?
12. If 6 cows cost $234, what will 1 cow cost?
13. How many eggs must I put into each of 5 baskets that I may put 480 eggs in all?
14. If a man earns $128 in 4 weeks, how much does he earn in 1 week?
15. There are 168 hours in one week; how many hours are there in one day?
16. If 1 man can do a piece of work in 186 hours, in how many hours can 6 men do it?
17. How many Sundays are there in 728 days?
18. If $117 is paid for 9 weeks' board, what must be paid for 1 week's board?
19. A man divided $15,000 equally among his children, 5 sons and 1 daughter; what was the share of each?

21. LONG DIVISION.

NOTE. In the preceding articles, division has been performed mostly in the mind, results only being expressed. When the divisor is a large number, it is generally more convenient to have the work expressed in full; the process is then called Long Division.

Ill. Ex. VII. Divide 5,682 by 13.

```
Operation.   Hunds. Tens. Units.
13 ) 5682 ( 4 3 7
     52 hund.
     ----
      48 tens.
      39 tens.
      ----
       92 units.
       91 units.
       ----
        1 rem.
```

In 56 (hundreds), there are 4 (hundred) 13's. We write a figure 4 for the first figure of the quotient; 4 (hundred) 13's equal 52 (hundred), which taken out of 56 (hundred), leaves 4 (hundred). Uniting the remainder with the next term of the dividend, we have 48 (tens), in 48 (tens), there are 3 (tens) of 13's; we write 3 for the second figure of the quotient; 3 (tens) of 13's equals 39 (tens), which taken out of 48 (tens), leaves 9 (tens). Uniting the remainder with the next term of the dividend, we have 92 units ; in 92 units there are seven 13's, and 1 remains; therefore, etc.

Ill. Ex. VIII. Divide 4,327 by 14.

```
Operation.
14 ) 4327 ( 309
     42
     ----
      127
      126
      ----
        1
```

Whenever the divisor is not contained in any partial dividend, put a zero in the expression for the quotient, bring down the next figure of the dividend, and divide.

1. Divide 3,680 by 13; by 14; by 15.
2. Divide 1,006 by 14; by 16; by 18.
3. Divide 3,060 by 21; by 32; by 153.
4. Divide 8,164 by 73; by 36; by 232.
5. Divide 180,000 by 54; by 82; by 176.
6. Divide 123, 456 by 98; by 76; by 543.

Ill. Ex. IX. Divide 485 by 100.

1|0 0) 4|8 5
Ans. 4 . . . 8 5 rem.

Whenever the divisor is a unit of any order, as 10, 100, 1,000, etc., we can express the quotient by cutting off as many of the right hand figures of the expression for dividend as the expression for divisor has zeros. The figures which are not cut off denote the quotient, and the figures cut off denote the remainder.

7. Divide 1,850 by 10; by 100; by 1,000.
8. Divide 45,867 by 10; by 100; by 1,000.
9. Divide 8,560 by 1,000; by 100; by 10.

Ill. Ex. X. Divide 748 by 30.

3|0) 7 4|8
Ans. 2 4 . . 2 8 rem.

Whenever the expression of the divisor has zeros at the right, cut them off, also cut off as many figures at the right of the expression of the dividend, and then divide. If there be a remainder after this division, at the right of the expression of the remainder, express the undivided part of the dividend; the true remainder will be thus expressed.

10. Divide 543 by 80; 3,654 by 240.
11. Divide 37,806 by 2,500; 59,004 by 3,000.
12. If 27 acres of land cost $945, what will 1 acre cost?
13. A farmer had 576 eggs to pack in baskets which held 48 eggs each; how many baskets did it require?
14. From Albany to Boston is 200 miles; how many hours would it take a locomotive to go that distance, at the rate of 40 miles an hour?
15. How many pieces of cloth are there in 535 yards, if there are 43 yards in each piece?
16. A barrel of flour contains 196 pounds; how many barrels can be filled from 5,836 pounds?
17. If in one hogshead there are 63 gallons, how many hogsheads are there in 81,045 gallons?
18. If a man saves $3 in 1 week, how many weeks will it take him to save $1,575? How many years will it take?

MISCELLANEOUS EXAMPLES.

22. 1. A man left $366,000 to his wife and 3 children: his wife was to have 1 half, and his children were to share the remainder equally; what was the share of each?

2. A man whose salary was $1,300, found at the end of the year that his expenses had been as follows: for groceries, $216; for milk, $18; meat, $83; wood, $13; coal, $89; clothing, $290; hired help, $156; medical service, $55; house rent, $225; taxes, $27; repairs and incidental expenses, $95; charity, $19. How much did he save during the year?

3. The product of two numbers is 3,705, one of the numbers is 57; what is the other number?

4. If the dividend is 1,312, and the divisor is 82, what is the quotient?

5. If the quotient is 43, and the dividend 1,333, what is the divisor?

6. If the quotient is 800, and the divisor 176, what is the dividend?

7. If the quotient is 37, the divisor 56, and the remainder 16, what is the dividend?

8. A merchant bought 200 barrels of apples at $8 a barrel: he sold 50 barrels at $11 a barrel; what must he receive a barrel for the remainder, that he may gain $450 on the whole?

9. The length of the Atlantic coast belonging to the United States is 6,186 miles; of the Pacific coast, 2,281 miles; of the shore of the Gulf of Mexico, 3,467 miles; of the shore of the Northern Lakes, 3,600 miles. How many days would it take to row this distance, at the rate of 25 miles a day?

10. How many days would it take a Laplander to travel in his sledge from the Arctic Ocean to the Black Sea, the distance being 1,700 miles, at the rate of 50 miles a day?

11. How many days would it take a vessel to go from Havana to New Orleans, the distance being 650 miles, at the rate of 200 miles a day?

REVIEW IN ADDITION.

An indefinite number of problems in Addition can be made from the following circle of figures : —

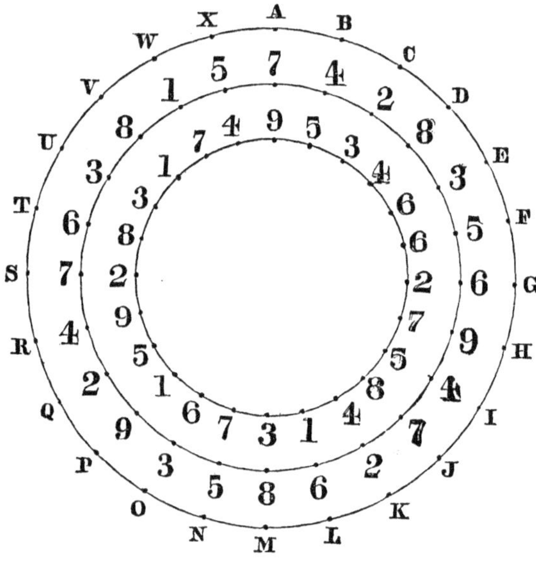

ILLUSTRATIONS.

1. Commencing at any letter in the outer circle, the pupil adds three, four, or more numbers towards the right or towards the left, as indicated by the teacher; thus, commencing at A, and adding towards the right, he says, 7, 11, 13, 21, 24, etc.; adding towards the left, he says, 7, 12, 13, 21, 24, 27, etc.

2. Commencing at a particular letter, the pupil adds the number against that letter in the outer circle to the successive numbers of the inner circle ; or, commencing at a number of the inner circle, he adds to it the successive numbers of the outer circle.

3. The teacher mentions some number, and the pupil adds to it the numbers of either circle, commencing at the different letters. For example, the teacher mentions 15 : the pupil commencing at A, in the outer circle, and adding towards the right, says, 15, 22, 26, 28, 36, etc.

REVIEW
IN
SUBTRACTION AND MULTIPLICATION.

Find the difference between the numbers in any two columns.							
* Multiply the numbers in one column by those in another.				Subtract from the following numbers the numbers from 2 to 12 inclusive.			
1st.	2d.	3d.	4th.	A.	B.	C.	D.
9	5	12	11	16	41	57	85
8	12	7	5	30	39	63	94
4	7	6	9	28	42	59	87
10	11	3	4	14	44	64	91
5	6	6	12	32	55	76	89
11	2	10	7	17	49	71	80
8	3	9	11	21	52	65	95
7	9	8	8	12	48	68	83
6	10	12	12	31	37	70	98
12	3	7	8	25	40	56	79
2	7	11	6	33	53	75	93
4	12	5	5	26	50	67	97
9	10	8	11	18	45	74	86
3	6	10	10	29	36	69	90
11	5	4	9	20	34	71	88
8	4	11	11	19	47	73	92
2	6	8	2	24	54	60	78
5	8	10	5	13	38	62	82
6	4	4	10	27	35	77	96
12	9	9	2	22	46	66	84
7	7	2	12	15	51	72	81
5	3	0	10	23	43	58	99

* Subtract the numbers in one column from those in another, *prefixing* the figure 1 to the number used as the minuend when it consists of a single figure, and *adding* 1 to the left hand figure when the number consists of two figures.

REVIEW IN DIVISION.

The following table is designed for a review in Division. Commencing at the top or bottom of any column, let the pupils divide the numbers, successively, in that column, by the numbers indicated at the top of the column. Thus, in column A, commencing at the top, 3 in 26, 8 times, and 2 remain, or $8\frac{2}{3}$ times; 3 in 15, 5 times, etc.

Divide by the Numbers from					Divide
3 to 12.*	5 to 12.*	6 to 12.*	8 to 12.*	10 to 12.*	by 12.
A.	B.	C.	D.	E.	F.
26	39	63	87	108	135
15	44	59	91	115	141
31	49	71	80	105	127
23	52	68	98	112	133
12	37	75	79	120	142
29	50	67	86	116	138
30	36	69	78	109	143
27	54	73	92	118	136
19	35	62	84	101	129
22	43	77	96	113	140
16	51	66	81	119	132
24	46	58	99	110	123
17	38	64	82	102	130
20	47	72	90	117	126
32	34	61	88	103	134
13	45	74	93	121	144
18	53	56	85	114	128
33	40	70	97	122	137
25	48	65	83	104	124
28	55	76	95	111	131
14	42	64	89	107	139
21	41	57	94	106	125

* Inclusive.

www.ingramcontent.com/pod-product-compliance
Lightning Source LLC
LaVergne TN
LVHW011233110826
845150LV00006B/1630

9781425514037